SpringerBriefs in Meteorology

SpringerBriefs in Meteorology present concise summaries of cutting-edge research and practical applications. The series focuses on all aspects of meteorology including, but not exclusive to; tornadoes, thunderstorms, lightning, hail, rainfall, fog, extratropical and tropical cyclones, forecasting, snowfalls and blizzards, dust storms, clouds. The series also presents research and information on meteorological technologies, meteorological applications, meteorological forecasting and meteorological impacts (reports of notable worldwide weather events). Featuring compact volumes of 50 to 125 pages (approx. 20,000–70,000 words), the series covers a range of content from professional to academic such as: a timely reports of state-of-the art analytical techniques, literature reviews, in-depth case studies, bridges between new research results, snapshots of hot and/or emerging topics. Author Benefits: Books in this series will be published as part of Springer's eBook collection, with millions of users worldwide. In addition, Briefs will be available for individual print and electronic purchase. SpringerBriefs books are characterized by fast, global electronic dissemination and standard publishing contracts. Books in the program will benefit from easy-to-use manuscript preparation and formatting guidelines, and expedited production schedules. Both solicited and unsolicited manuscripts are considered for publication in this series. Projects will be submitted for editorial review by editorial advisory boards and/or publishing editors. If you are interested to submit a proposal, please contact your Publisher, Dr. Michael Leuchner.

More information about this series at http://www.springer.com/series/13553

Paolo Cristofanelli • Erika Brattich
Stefano Decesari • Tony Christian Landi
Michela Maione • Davide Putero
Laura Tositti • Paolo Bonasoni

High-Mountain Atmospheric Research

The Italian Mt. Cimone WMO/GAW Global Station (2165 m a.s.l.)

Springer

Paolo Cristofanelli
Institute for Atmospheric Sciences and Climate (ISAC)
National Research Council of Italy (CNR)
Bologna, Italy

Stefano Decesari
Institute for Atmospheric Sciences and Climate (ISAC)
National Research Council of Italy (CNR)
Bologna, Italy

Michela Maione
Institute for Atmospheric Sciences and Climate (ISAC)
National Research Council of Italy (CNR)
Bologna, Italy

Department of Pure and Applied Sciences
University of Urbino
Urbino, Italy

Laura Tositti
Department of Chemistry "G. Ciamician"
University of Bologna
Bologna, Italy

Erika Brattich
Department of Chemistry "G. Ciamician"
University of Bologna
Bologna, Italy

Tony Christian Landi
Institute for Atmospheric Sciences and Climate (ISAC)
National Research Council of Italy (CNR)
Bologna, Italy

Davide Putero
Institute for Atmospheric Sciences and Climate (ISAC)
National Research Council of Italy (CNR)
Bologna, Italy

Paolo Bonasoni
Institute for Atmospheric Sciences and Climate (ISAC)
National Research Council of Italy (CNR)
Bologna, Italy

ISSN 2199-9112 ISSN 2199-9120 (electronic)
SpringerBriefs in Meteorology
ISBN 978-3-319-61126-6 ISBN 978-3-319-61127-3 (eBook)
DOI 10.1007/978-3-319-61127-3

Library of Congress Control Number: 2017945270

Printed on acid-free paper

This Springer imprint is published by Springer Nature
The registered company is Springer International Publishing AG
The registered company address is: Gewerbestrasse 11, 6330 Cham, Switzerland

Contents

The "O. Vittori" Observatory at Mt. Cimone: A "Lighthouse" for the Mediterranean Troposphere

Abstract The Mediterranean basin represents a global hotspot for climate change, air quality, and anthropogenic contributions to these issues. Since the early 1990s at Mt. Cimone, the highest peak of Italian northern Apennines, an observatory is performing continuous measurements of atmospheric composition. The Italian climate observatory "O. Vittori" is a research infrastructure managed by the Institute for Atmospheric Sciences and Climate (ISAC) of the National Research Council of Italy (CNR) and hosted by the Italian Air Force. It is part of the WMO/GAW global station "Mt. Cimone" (GAW id: CMN). Due to the completely free horizon, high altitude, and great distance from major pollution sources, CMN represents a strategic platform to study the chemical–physical characteristics and climatology of the free troposphere in the South Europe and Mediterranean basin. At this observatory, continuous monitoring of climate-altering compounds (trace gases and aerosol), solar radiation, as well as meteorological parameters is carried out. Besides providing a historical perspective of scientific research at CMN, we characterize the measurement site, and we describe the current observatory technical facilities, including the e-access to data and the services for near-real-time data delivery. Good practices for educational and outreach activities are also presented.

Keywords Mediterranean basin • Monte Cimone • Atmospheric composition • Aerosol particles • Trace gases • Meteorological parameters • Open data • Outreaching

1 Introduction

Southern Europe and Mediterranean basin are recognized as hotspot regions both in terms of climate change (e.g., Giorgi and Lionello 2008) and air quality (Monks et al. 2009). Meteorological conditions characterized by frequent clear sky and high solar radiation favor the occurrence of ozone (O_3) photochemical production also, thanks to the availability of natural and anthropogenic precursors. In particular, large amounts of anthropogenic pollutants emitted in continental Europe are transported toward the Mediterranean free troposphere and boundary layer (e.g., Safieddine et al. 2014). Saharan dust outbreaks from northern Africa (Querol et al.

P. Cristofanelli et al., *High-Mountain Atmospheric Research*, SpringerBriefs in Meteorology, DOI 10.1007/978-3-319-61127-3_1

Table 1 List observation programs running at the "O. Vittori" observatory at Mt. Cimone (Uniurb: Urbino University; Unibo: Bologna University)

Observations	Starting year	Lead institution
Surface O_3	1996	CNR-ISAC
NO, NO_2	2012	CNR-ISAC
SO_2	2014	CNR-ISAC
CO	2008	CNR-ISAC/Uniurb
CH_4	2008	Uniurb
N_2O	2008	Uniurb
SF_6	2008	Uniurb
CFCs, HCFCs	2002	Uniurb
HFCs	2002	Uniurb
Columnar NO_2	1993	CNR-ISAC
Aerosol size distribution (10 – 500 nm)	2005	CNR-ISAC
Aerosol size distribution (300 nm – 10 μm)	2000	CNR-ISAC
Aerosol scattering	2005	CNR-ISAC
Aerosol absorption	2005	CNR-ISAC
Equivalent BC	2005	CNR-ISAC
Aerosol chemistry (PM_1 – PM_{10})	2005	CNR-ISAC
Natural radionuclides (^{7}Be, ^{210}Pb, ^{222}Rn)	1998	Unibo
Solar photometry	2016	CNR-ISAC
Meteorological parameters and solar radiation	1996	CNR-ISAC

2009) and open biomass burning (Turquety et al. 2014) further exacerbate air quality and the influence of anthropogenic emissions on the regional climate (Mallet et al. 2013). Water scarcity, the concentration of economic activities in coastal areas, and reliance on climate-sensitive agriculture together with demographic, social, cultural, economic, and environmental changes are other critical factors which make this region particularly exposed to climate change and air quality worsening.

For these reasons, it is important to have long-term observations of essential climate variables with well-assessed quality. In particular, in situ or ground-based observations at mountain sites can provide key information about background atmospheric composition. They also provide the opportunity to investigate and assess the impact of natural and anthropogenic-related processes to atmospheric variables and then to regional climate and air quality.

Since the early 1990s of the last century, continuous observations of atmospheric composition have been carried out in Italy at Mt. Cimone (44°12' N, 10°42′ E, 2165 m a.s.l.), by means of the "O. Vittori" atmospheric observatory managed by the National Research Council of Italy (CNR). Nowadays, at this locations, thanks to the collaboration with different research institutions (Italian Air Force - Centro Aeronautica Militare di Montagna, and Urbino University), several scientific programs covering research on reactive and greenhouse gases, chemical and physical aerosol particle properties, meteorological parameters, and solar radiation (Table 1) are carried out.

Fig. 1 Mt. Cimone peak, with the observatories of Italian Air Force and CNR

2 Presentation of the Measurement Site

The "O. Vittori" observatory is located at the top of Mt. Cimone (44°12′ N, 10°42′ E, 2165 m a.s.l., Fig. 1), the highest peak of the Northern Apennines, at the border line of two different climatic regions: the continental Europe northward and the Mediterranean basin southward.

The Mediterranean Sea is about 50 km to the SW of the measurement site (Fig. 2). The Po Valley, one of the most polluted region in Europe (Monks et al. 2009), lies immediately North of the site. The closest inhabited areas are small villages (1500 inhabitants) placed 15 km from and about 1100 m below the observatory, whereas major towns (500,000 inhabitants) are situated in the lowlands about 60 km away (Bologna, Firenze). Mt. Cimone is characterized by a 360° free horizon which allows the air masses to reach the measurement site without any topographic channeling. As assessed by the GEOMON EU Project (Henne et al. 2010), Mt. Cimone is characterized by a wide catchment area for air masses older than 48 h which experienced interaction with terrain surface, making this measurement site well representative of the atmospheric air masses originating within the central Mediterranean basin (Fig. 2). Within several kilometers from the site, there are no crops, and human activity is very limited. This makes the measurement site very suitable for investigating the background conditions of the Mediterranean troposphere as well as the direct impact of surface emissions to them.

The "O. Vittori" observatory is located above the timberline, and only some patches of grass can be found on the mountaintop, which is mostly rocky and covered with snow for 6–7 months a year. In spring, the last snowfall generally takes place in mid-April, while in autumn the early snowfall usually occurs in mid-November.

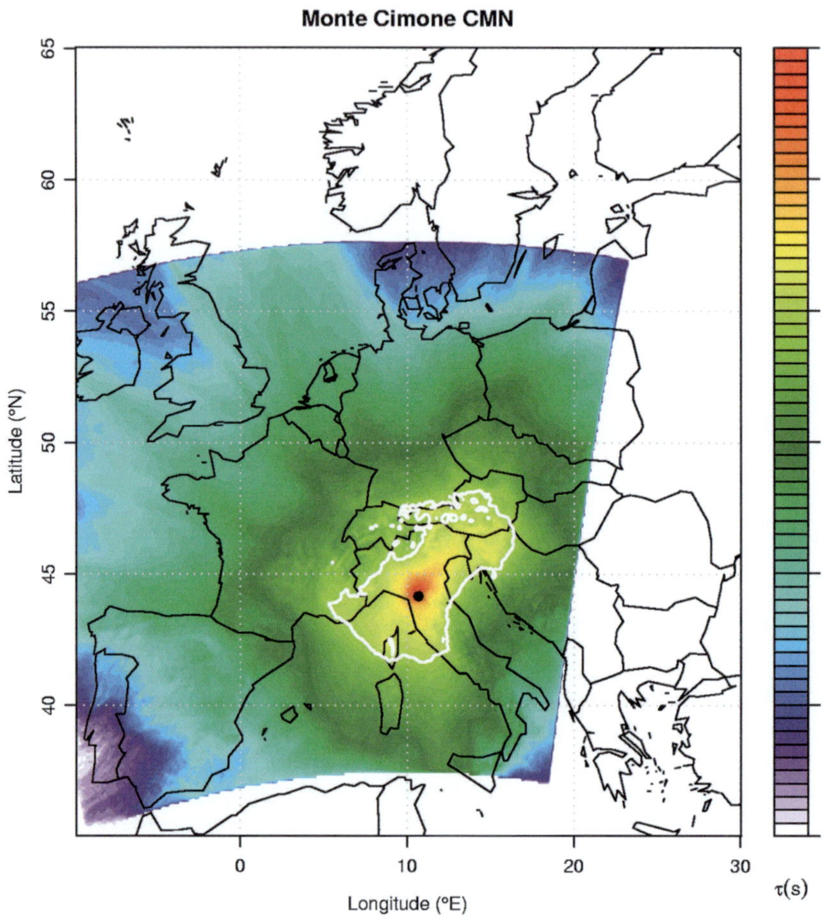

Fig. 2 Surface footprint 48 h catchment area for CMN (*white contour*). The catchment area is given by the intersection of the topography with the volume containing the largest residence time densities and comprises 50% of the total residence time. FLEXPART model (0.5° × 0.5° horizontal resolution) was used to derive catchment areas (From www.geomon.empa.ch)

Colombo et al. (2000) provided a meteorological description of Mt. Cimone weather regime from 1946 to 1999. They showed that Mt. Cimone is the windiest meteorological station in Italy (wind speed higher than 200 km/h is occasionally observed) and the prevailing winds blow from S–SW and N–NE. This was evident also by analyzing the wind observations carried out from 1996 to 2015 at the "O. Vittori" observatory (Fig. 3). In winter and fall, when temperate air masses flow from the Mediterranean Sea, the temperature only occasionally lies above 0 °C. On the other hand, when continental air comes from N and NE, the temperature is usually several degrees below 0 °C. As deduced by "O. Vittori" observatory measurements, the lowest monthly mean temperature was in February 2005 (−7.9 °C) and

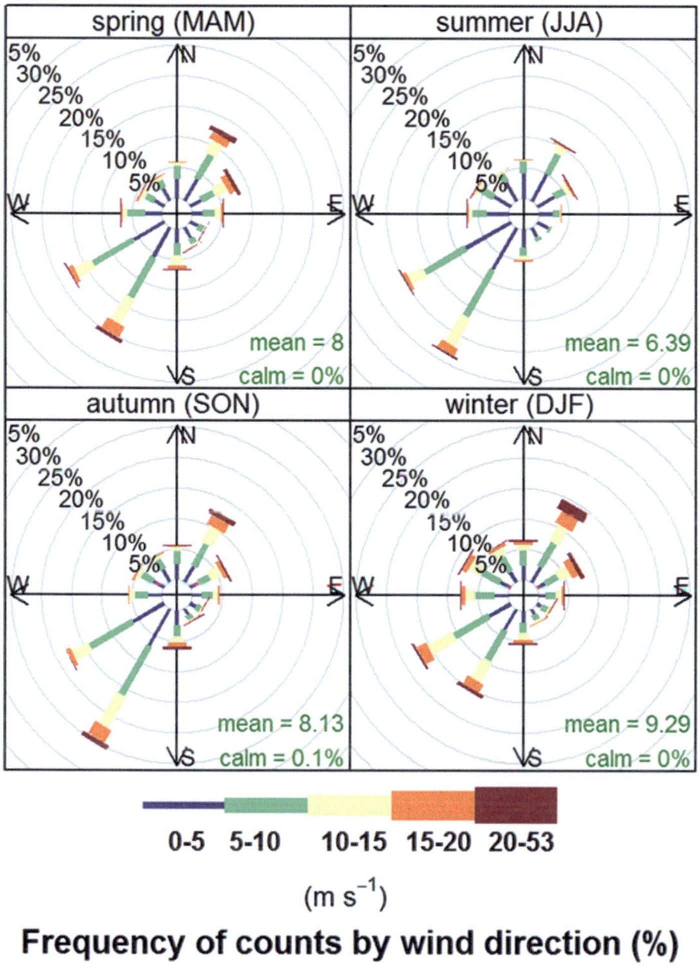

Fig. 3 Wind speed/direction frequencies by season at CMN (1996–2015). Wind speeds are split into the intervals shown by the scale in each panel. The *grey circles* show the 5% frequency intervals

the highest in July 2015 (14.9 °C). The annual mean (as deduced by averaging monthly mean values) is 2.1 °C. As reported by previous investigations, the atmospheric observations carried out at CMN can be considered representative of the free tropospheric conditions of the Mediterranean basin/South Europe during the cold months (see, e.g., Bonasoni et al. 2000; Henne et al., 2010) as well as during nighttime in the warm season. However, from April to September, the measurement site can be affected by "thermal" wind circulation and convective vertical transport of air masses. Indeed, during daytime, upslope and valley winds together with diurnal planetary boundary layer (PBL) growth and entrainment processes can favor the vertical transport of polluted PBL air to the measurement site (Colombo et al. 2000; Cristofanelli et al. 2009).

3 History

Across the last five centuries, Mt. Cimone represented a landmark for scientific researches.

In 1655, Fathers Riccioli and Grimaldi made the first measurement of the height of Mt. Cimone. In 1671, for the first time in Italy, the mathematician Geminiano Montanari went to Mt. Cimone to use a barometer for altitude measurement. In 1896, military engineers from Modena Duchy built a small pyramid on the top of the mountain for a survey to write the Duchy's new map. In 1817, Dr. Brioschi, director of the Naples observatory, performed geographical studies at Mt. Cimone to assess the level of Tyrrhenian and Adriatic Seas. In 1823, Father Inghirani used the small pyramid on the top of Mt. Cimone for triangulation studies of the Tuscan Grand Duchy. In 1888, a 14 m tower was built, dedicated to the memory of Geminiano Montanari and assigned to operate as scientific observatory and mountain cabin. It was used to perform earlier studies on electrical discharges, lightning rods, and solar radiation. Due to the harsh weather conditions affecting Mt. Cimone, the tower experienced strong damages across the years, and it was completely knocked down in the 1930s.

In 1936, the Air Force Ministry built small barracks on the top for meteorological observations and flight assistance: this represents the first presence of the Italian Air Force at Mt. Cimone. In 1937, the Italian Air Force inaugurated a Meteorological Station at the mountain top, and after the World War II (1948), the Osservatorio Scientifico Sperimentale di Meteorologia Aeronautica (OSSMA) "Air Force Experimental Scientific Meteorological Observatory" was built. This scientific observatory was used to perform atmospheric investigations about condensation nuclei by Dr. D. Fedele and Prof. O. Vittori. From 1950 to 1959, several research activities were carried out by the Italian Air Force at Mt. Cimone: studies about thunderstorms, hail, fog formation, and environmental radioactivity. In 1964, the first studies about atmospheric pollution were started, while in 1979 the monitoring program for atmospheric CO_2 measurement (still active at the Italian Air Force Meteorological Observatory, representing one of the longest time series of CO_2 observations in background conditions over the European continent; see Fig. 4) was implemented.

In 1981, Ottavio Vittori, who had already been the commanding officer of the Air Force Military Mountain Centre (CAMM) in Sestola and subsequently director of the CNR-FISBAT in Bologna, drew up a contract with the Italian Air Force to allow the CNR for using the "Romualdi" mountain cabin for scientific purposes. From this year, the CNR executed measurements of atmospheric physics and chemistry at Mt. Cimone.

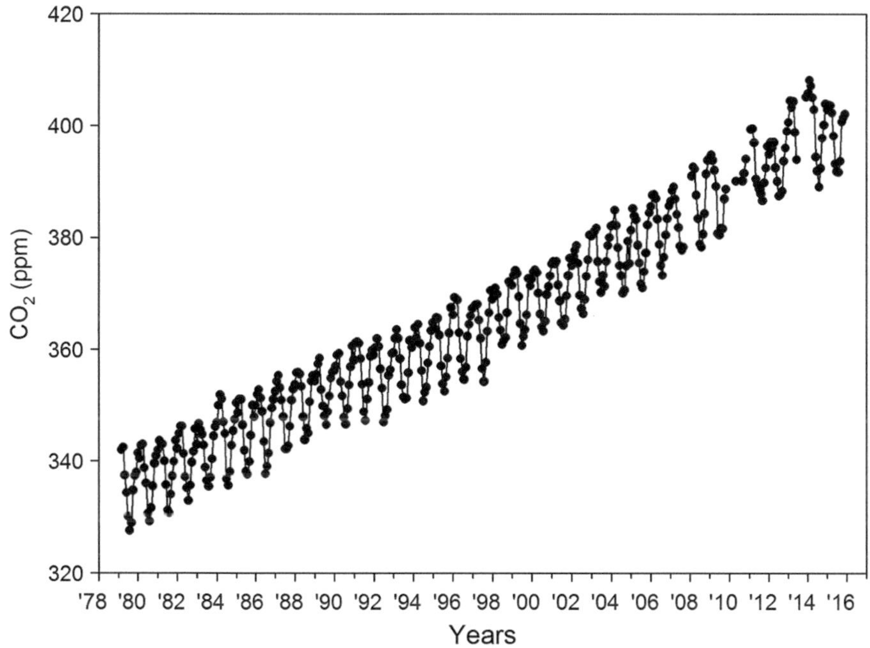

Fig. 4 Long-term (1979–2016) CO_2 time series at CMN (Data by Italian Air Force by way of GAW-WDCGG, http://ds.data.jma.go.jp/)

Starting from 1991, measurements of atmospheric compounds started being continuously carried out, and in 1993 a Differential Optical Absorption Spectrometry (DOAS) system for the measurements of stratospheric NO_2 column amount was installed at the station (see Chap. 3). In 1996, under the framework of the VOTALP EU Project, surface ozone, carbon monoxide, and natural radioactivity continuous measurements started at the CNR observatory. In July 1998, under the management of Paolo Bonasoni, the CNR observatory at Mt. Cimone is called after Prof. Ottavio Vittori. From 1996 to 2000, CNR refitted the "O. Vittori" observatory (Fig. 5) and equipped it with the most advanced devices for defenses against lightning and ice rimming to allow continuous monitoring of atmospheric composition.

4 The Infrastructure

The CNR "O. Vittori" observatory (Fig. 5) is hosted by the buildings of the Italian Air Force operative base on the top of Mt. Cimone. In particular, it is held in a former mountain cabin built in 1939 by the Italian Alpine Club (CAI) and called after "Gino Romualdi," an Italian soldier who died during the World War I. In 1943, during the World War II, the cabin was destroyed by military activities and then rebuilt

Fig. 5 External view of the "O. Vittori" observatory (Picture by P. Bonasoni)

in 1947. Later (1949), the "Romualdi" cabin was confiscated by the Italian Air Force which used it for military purposes.

The "O. Vittori" observatory can be reached from Sestola (Modena district) and the Ninfa Lake by a military road ending at 1850 m a.s.l. From there, the facilities of the Italian Air Force operative base (located inside the mountain) allow to reach the research infrastructures located at the top of the mountain.

4.1 Laboratories and Facilities

Currently, the CNR observatory offers the possibility to use an equipped infrastructure to make accurate investigations of the processes affecting atmospheric background composition. The station is completely automated, and no scientific personnel is working permanently at the site. It is composed by the "reactive gas" laboratory (first floor, Fig. 6), the "aerosol" laboratory (ground floor), and the "greenhouse gas" laboratory (ground floor). Two specific designed intakes for trace gas and aerosol particles allow continuous air sampling even during extreme weather conditions.

The trace gas sampling system (which drain air toward the reactive and greenhouse gas laboratories) is composed by an intake line extending 2 m above the roof

Fig. 6 View of the laboratory for "reactive gases" at the "O. Vittori" observatory

and 7 m above the ground and consists of a glass tube through which the sampled air is passed at a high flow rate to minimize the permanence time (less than 5 s). The parts of the system that are outside the building are heated to prevent ice formation during the cold season. Inside the building, the analyzers draw their sample air throughout Teflon and stainless steel lines.

The aerosol sampling system (Fig. 7) is composed by a stainless steel tube extending for 3 m above the terrace of the observatory (5.5 m from the terrain). Following the EUSAAR/ACTRIS/GAW guidelines, a laminar flow is a guarantee during the sampling (sampling flow is taken constant at 120 l min^{-1}). The air intake is equipped with a heated (to avoid ice rims) total size particle (TSP) sampling head. Internal RH and temperature are continuously monitored to have information about the sampling conditions. Since the ambient temperature of aerosol laboratory is higher than external temperature, water vapor condensation is avoided during the sampling, and only the "dry" fraction of aerosol is considered for analysis.

Two high-volume samplers for PM-10 and PM-10/PM-1 are accommodated in a specific designed shelter on the observatory terrace. The terrace (about 40 m^2) is equipped for hosting external experimental activity (e.g., solar photometry), and a small chemistry laboratory permits a clean treatment of collected samplings. Two

Fig. 7 External view of the sampling system for aerosol particles on the "O. Vittori" observatory terrace. The Apuan Alps can be seen in the background

bedrooms (six beds), toilettes, and a small kitchen allow for long stays at the station. This makes possible for scientists working at the observatory to continuously look after complex instrumentation during intensive field campaigns.

The observatory runs by electric energy (220 V AC, 50 Hz, 18 kW), and a set of UPS operates in emergency conditions for about 36 hours (depending on power consumption). Continuous fast internet connection is available, thanks to a fast internet connection provided by the Emilia-Romagna region throughout Lepida SpA. This high-speed internet connection allows the remote control of instrumentation as well as near-real-time (NRT) data delivery. Indeed, data from acquisition systems are mirrored to a station server and then delivered in NRT mode to CNR HQ in Bologna. A dedicated Web page reports the last 24 hours of online observations (www.isac.cnr.it/cimone/real-time).

4.2 *Open Access to Data*

In the recent years, the need for access to quality-assessed data about atmospheric composition increased notably. For these reason, CNR-ISAC recognized the importance of providing an open and free access to the historical time series of meteorological and atmospheric composition data recorded at the "O. Vittori" observatory. For these reasons, the system MonteCimone On-line VIsualization and Data Analyses (MOVIDA) was implemented under the Creative Commons License 3.0 (Fig. 8). It was launched on 5 July 2016 for celebrating the World Environment Day. The MOVIDA system, implemented in collaboration with Arpae Emilia-Romagna (SIMC), can be accessed at the web address http://www.isac.cnr.it/cimone/

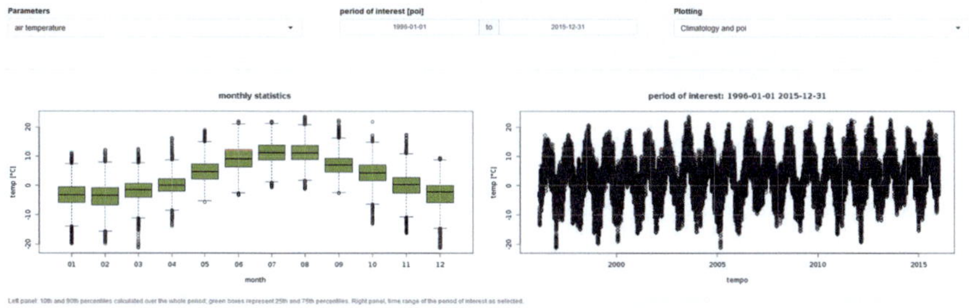

Fig. 8 Web interface of MOVIDA system, available at the web page http://www.isac.cnr.it/cimone/data-access

data-access. External users can access and download time series (with hourly time resolution) of meteorological parameter (air temperature, relative humidity, atmospheric pressure, wind speed, and direction), trace gases (O_3, CO, NO, NO_2, SO_2, CH_4, N_2O, SF_6, HFC134a, CFC-11, CFC-114), and aerosol properties (fine and coarse size distribution, absorption coefficient).

Moreover, MOVIDA represents a web resource by which external users can run some basic statistical analyses on CMN time series for each presented parameter and over flexible period of interests (see also Appendix A). Thanks to the packages Shiny (shiny.rstudio.com) and Openair (Carslaw and Ropkins 2012), it is possible to plot raw data, average (on different time scales: daily, monthly, and yearly), as well as to perform smoothing of time series to obtain information about long-term tendencies.

5 Educational Activities

Due to the wide and direct impacts of atmospheric composition variability to the society (e.g., population health, cultural heritage) and environment (e.g., loss of biodiversity), CNR-ISAC launched several outreaching and educational activities to increase the public awareness on these topics. In this regard, the "O. Vittori" observatory represents a unique platform to introduce school pupils and generic public to atmospheric research and to disseminate the most recent knowledge about impacts of natural and anthropogenic-related processes to climate change and air quality. To these aims, a challenging educational project has been launched on 2004 by CNR-ISAC with the collaboration of Italian Air Force and Frignano Regional Park: "Il Sentiero dell'Atmosfera" ("A footpath to atmosphere"; see www.sentieroatmosfera.it). A mountain footpath, accessible from June to October, starting at 1850 m a.s.l. and reaching the top of Mt. Cimone was equipped with 14 panels presenting and discussing the research activities carried out at the observatory (Fig. 9). After the trip to the top of Mt. Cimone, people are allowed to visit the "O. Vittori" observatory under the tutoring of CNR-ISAC researchers and technicians to learn "where"

Fig. 9 Generic public participating to the "Sentiero dell'Atmosfera" experience (Picture by P. Bonasoni)

and "how" measurements concerning the atmosphere and the Earth's climate are carried out by scientific community.

This also represents a great opportunity for scientists to directly interact with society and to disseminate correct information about the state of our atmosphere and climate, about the impact of anthropogenic-related processes, and about the impacts of atmospheric composition variability to population health, wellness, and environment. Since 2004, thousands of people (school pupils and generic public) participated to this project.

6 Future Perspectives

Recent scientific "mysteries" pointed out the importance of having long-term ground-based measurements of atmospheric composition even if the "era" of satellite and earth–system model: for example, the increase of CH_4 global levels from 2007, the still not clearly declining average O_3 levels despite the wide precursor emission over North America and Europe in the last decades, and the discrepancies between bottom-up and top-down inventories for specific climate forcers (as an instance for CH_4 and carbon tetrachloride).

Moreover, high-quality, continuous observation activities like those carried out at CMN can profitably contribute to *science services*. These *services* include the dissemination to local/regional agencies or companies of state-of-the-art knowledge about technical solution or strategies for the environmental monitoring and the data analysis/integration, the implementation of near-real-time services for data delivery/data quality assessment, and the early warning activities (see the case of the eruption of the Eyjafjallajökull in 2010, as described in Chap. 5). As will be shown in Chap. 2, the observations carried out at CMN can be used to assess the variability of anthropogenic and natural emissions of climate-altering or pollutant species in supporting national agencies or governments for the verification of emission inventories or for assessing the efficiency of mitigation measures. This will represent an important task in the next decades, when national governments must respond to international protocols for the decrease of emissions of climate-altering/pollutant species. Moreover, by representing an attractive location for "scientific tourism," research infrastructure like CMN can also contribute in the increasing public awareness about different themes of atmospheric sciences. This can be realized by specific educational or communication activities like those described in Sect. 4 devoted to reporting updated knowledge about scientific evidences and facts, communicating uncertainties, and suggesting good practices to be adopted by citizens.

The "road map" for the scientific observations at CMN for the next decade includes the participation to European Environmental Research Infrastructures, to further foster, over a long-term framework, the integration of the CMN atmospheric composition observations in the European and global perspectives. In this context, an implementation of the profiling capability at CMN will be pursued in the next years by upgrading of the current DOAS system (see Chap. 3) with the multi axis DOAS (MAXDOAS) technique and by the installation of an automated aerosol LiDaR.

Of course, for the achievement of these long-term goals, it is paramount to have adequate plans and strategies for hiring technical and scientific personnel. In this perspective, a continuative support from national and regional governments will be necessary, to guarantee the execution of high-quality observations and the production of high-quality science over the long term.

7 Conclusions

The "O. Vittori" observatory is a research infrastructure managed by the Institute of Atmospheric Sciences and Climate (ISAC) of the National Research Council (CNR). It is the only high mountain station for atmospheric research in both South of the Alps and the Po basin, and it represents a strategic platform to study the chemical–physical characteristics and climatology of the South Europe and North Mediterranean basin. Together with the Meteorological observatory of the Italian Air Force, the "O. Vittori" observatory is the only "Global Station" in Italy and over the Mediterranean basin within the Global Atmosphere Watch (GAW) program by the World Meteorological Organization (WMO). Thanks to its technical facilities and the accessibility provided year-round by the Italian Air Force, the "O. Vittori" observatory is a key location where investigation of atmospheric processes is performed to disseminate updated knowledge about the current status of the Earth's atmosphere and climate.

In the chapters of this book, we present more than 20 years of atmospheric composition research studies carried out at this observatory. We will provide a comprehensive and updated result about the main research topics carried out at the "O. Vittori" observatory: non-CO_2 greenhouse gases (Chap. 2), reactive gases (Chap. 3), environmental radiochemistry (Chap. 4), and aerosol properties (Chap. 5).

Acknowledgments CNR-ISAC gratefully thanks the Italian Air Force for providing access to the operational base at the top of Mt. Cimone and to the "O. Vittori" Italian Climate Observatory. We also gratefully thank all the people that along the years made possible the execution of atmospheric research at the Observatory: Cesarino Magera (1930–†2016), Pio Giambi, Ferrero Nizzi, Paolo and Domenico Amidei, Nicola Gherardini and the staff at CNR-ISAC (Francescopiero Calzolari, Fabrizio Roccato, Angela Marinoni, Ubaldo Bonafè, Franco Evangelisti) as well as Maurizio Caleca and Lucio Altieri (CNR building heritage office). A special mention is given to Dr. Claudio Tomasi, former director of CNR-ISAC, who strongly supported the start of the "O. Vittori" observatory activities and to Dr. Tiziano Colombo, former director of the CAMM Monte Cimone.

References

Bonasoni P, Stohl A, Cristofanelli P, Calzolari F, Colombo T, Evangelisti F (2000) Background ozone variations at Mt. Cimone Station. Atmos Environ 34:5183–5189

Carslaw DC, Ropkins K (2012) Openair – an R package for air quality data analysis. Environ Model Softw 27–28:52–61

Colombo T, Santaguida R, Capasso A et al (2000) Biospheric influence on carbon dioxide measurements in Italy. Atmos Environ 34:4963–4969

Cristofanelli P, Marinoni A, Arduini J, Bonafè U, Calzolari F, Colombo T, Decesari S, Duchi R, Facchini MC, Fierli F, Finessi E, Maione M, Chiari M, Calzolai G, Messina P, Orlandi E, Roccato F, Bonasoni P (2009) Significant variations of trace gas composition and aerosol properties at Mt. Cimone during air mass transport from North Africa – contributions from wildfire emissions and mineral dust. Atmos Chem Phys 9:4603–4619. doi:10.5194/acp-9-4603-2009

Giorgi F, Lionello P (2008) Climate change projections for the Mediterranean region. Glob Planet Chang 63:90–104

Henne S, Brunner D, Folini D, Solberg S, Klausen J, Buchmann B (2010) Assessment of parameters describing representativeness of air quality in-situ measurement sites. Atmos Chem Phys 10:3561–3581. doi:10.5194/acp-10-3561-2010

Mallet M, Dubovik O, Nabat P et al (2013) Absorption properties of Mediterranean aerosols obtained from multi-year ground-based remote sensing observations. Atmos Chem Phys. doi:10.5194/acp-13-9195-2013

Monks PS, Granier C, Fuzzi S et al (2009) Atmospheric composition change – global and regional air quality. Atmos Environ 43:5268–5350

Querol X, Alastuey A, Pey J et al (2009) Variability in regional background aerosols within the Mediterranean. Atmos Chem Phys. doi:10.5194/acp-9-4575-2009

Safieddine S, Boynard A, Coheur P-F, Hurtmans D, Pfister G, Quennehen B, Thomas JL, Raut J-C, Law KS, Klimont Z, Hadji-Lazaro J, George M, Clerbaux C (2014) Summertime tropospheric ozone assessment over the Mediterranean region using the thermal infrared IASI/MetOp sounder and the WRF-Chem model. Atmos Chem Phys 14:10119–10131. doi:10.5194/acp-14-10119-2014

Turquety S, Menut L, Bessagnet B, Anav A, Viovy N, Maignan F, Wooster M (2014) APIFLAME v1.0: high-resolution fire emission model and application to the Euro-Mediterranean region. Geosci Model Dev. doi:10.5194/gmd-7-587-2014

Non-CO_2 Greenhouse Gases

Abstract In the past decades, accurate and precise atmospheric measurements of radiatively active gases have been crucial in revealing the rapid and unceasing growth of their global concentrations that has been long recognized as the main driver of climate change. Even if carbon dioxide is the major anthropogenic contributor to radiative forcing, other gases such as methane, nitrous oxide, and halogenated gases are extremely relevant in climate issues because of their very high global warming potential. Continuous measurement programs of these gases are carried out at CMN in the frame of important international programs. This chapter reports an overview of the scientific results obtained based on 15 years of atmospheric measurements. These results have been achieved combining atmospheric data with different modeling approaches, with the aim of understanding the budget of these radiatively active gases and providing emission estimates at the regional scale. Such estimates constitute an important support to improve bottom-up emission data that each country is required to submit every year in the frame of the most important international global protocols aimed at combatting climate change.

Keywords Ozone depleting substances • Non-CO_2 greenhouse gases • Emission inventories

1 Introduction

Radiatively active gases able to absorb the infrared radiation emitted by the Earth's surface, thus altering its radiative balance, are known as greenhouse gases (GHGs). Many of them occur naturally in the atmosphere: carbon dioxide (CO_2), methane (CH_4), and nitrous oxide (N_2O). Others are synthetic: chlorofluorocarbons (CFCs), hydrofluorocarbons (HFCs), perfluorocarbons (PFCs), and sulfur hexafluoride (SF_6). They are also referred as well-mixed greenhouse gases (WMGHGs), being characterized by lifetimes long enough to be relatively homogeneously mixed in the troposphere (IPCC 2001). The increase of their atmospheric concentrations has been long recognized as the main driver of climate change (IPCC 1990).

The increase of atmospheric levels of CO_2, the most abundant GHG, is the major anthropogenic contributor to radiative forcing. However, because of the very high

P. Cristofanelli et al., *High-Mountain Atmospheric Research*, SpringerBriefs in Meteorology, DOI 10.1007/978-3-319-61127-3_2

global warming potential (GWP) values characterizing these compounds, the role of non-CO_2 GHGs, i.e., methane, nitrous oxide, and halogenated gases, is extremely relevant, accounting by more than the 44% of the total radiative forcing due to WMGHGs. The GWP of a GHG is the measure of how much energy the emissions of 1 ton of a gas will absorb over a given period of time, relative to the emissions of 1 ton of CO_2. Therefore, these gases have more significant climate change effects than CO_2 on a per ton basis. In addition, since many of them have shorter lifetimes with respect to CO_2, their reduction has greater short-term impacts. The potential climate benefits of reducing short-lived climate pollutant (SLCP) emissions have been recognized by the United Nations Environment Programme–Climate and Clean Air Coalition (UNEP-CCAC), an international initiative aiming at catalyzing rapid reductions in SLCPs, based on the assumption that their timely reduction can slow the rate of climate change within the first half of this century, i.e., before the long-lived CO_2 reduction measures will take effect (Velders et al. 2009; UNEP 2011; Shindell et al. 2012; Xu et al. 2013).

In the past decades, accurate and precise GHG atmospheric measurements have been crucial in revealing the rapid and unceasing growth of their global concentrations, which is the direct consequence of the rise of anthropogenic emissions mainly ascribable to the increasing energy demand of the growing world population.

Concern about the impacts of such increase led in 1992 to the stipulation of the United Nations Framework Convention on Climate Change (UNFCCC), whose ultimate objective is to achieve a stabilization of GHG concentrations in the atmosphere at a level that would prevent dangerous anthropogenic interference with the climate system.

Parties to the UNFCCC have the responsibility to report anthropogenic GHG emissions and sinks. These data are estimated and reported according to guidelines issued by the Intergovernmental Panel on Climate Change (IPCC) Task Force on National GHG Inventories (TFI) (IPCC 2006). In general, emission data are produced combining source-specific emission factors with activity data, following the so-called "bottom-up" approach. Such estimates are accurate when well-developed statistical systems are in place. Emissions from heterogeneous and dispersed sources are more difficult to be quantified. However, estimates can be improved, thanks to the availability of sophisticated methods and models, providing data that can be used as inputs to enhance activity data and emission factors. This independent approach, known as "top-down," is based on a combination of direct atmospheric measurements with inverse modeling methods (see, e.g., Bergamaschi et al. 2004; Nisbet and Weiss 2010; Weiss and Prinn 2011).

In the past decade, the top-down approach has gained recognition, thanks to the increased capability of producing high-quality atmospheric data and to the rapid development of inverse modeling techniques (Yokouchi et al. 2006; Greally et al. 2007; Stemmler et al. 2007; Meirink et al. 2008; Stohl et al. 2009; Kim et al. 2010; Miller et al. 2010; Montzka et al. 2010; Li et al. 2010; Stohl et al. 2010; Manning et al. 2011; Vollmer et al. 2011; Brunner et al. 2012; Keller et al. 2012; Yao et al. 2012; Houweling et al. 2014; Maione et al. 2014; O'Doherty et al. 2014; Hu et al. 2015; Lunt et al. 2015; Graziosi et al. 2015, 2016; Henne et al. 2016; Fang et al. 2016; Simmonds et al. 2016).

Long-term and/or continuous observations of the atmospheric levels of a large suite of non-CO_2GHGs are carried out in the frame of international programs like Advanced Global Atmospheric Gases Experiment (AGAGE, http://agage.mit.edu/), National Oceanic and Atmospheric Administration-Earth System Research Laboratory-Global Monitoring Division (NOAA-ESRL-GMD, https://www.esrl.noaa.gov/gmd/), Japanese National Institute for Environmental Studies (NIES), and University of California Irvine (UCI). Data from all these programs contribute to obtaining global abundances and trends of a number of species, as well as to evaluating their atmospheric lifetimes and emissions on a global scale. In addition, measurements conducted at sites not too far from major emission sources are also useful for increasing the resolution of the emission patterns from the global to the regional scale.

Monte Cimone (CMN), due to its location, is well suited for long-term monitoring of trace gas concentration trends in the free troposphere. In addition, the proximity to important anthropogenic source regions in central Europe makes it appropriate for regional source allocation studies. Continuous measurement programs of non-CO_2 GHGs at CMN started in 2001 in the frame of the European funded project System for Observations of halogenated Greenhouse in Europe (SOGE), which included four European sites: CMN, Jungfraujoch (JFJ, Switzerland), Mace Head (MHD, Ireland), and Zeppelin (ZEP, Norway). SOGE was aimed at establishing a fully inter-calibrated measurement network for halogenated compounds with the aim of deriving long-term trends as well as emissions on the regional European scale. Thanks to the involvement of MHD station, the SOGE network has been directly linked to the global AGAGE network and subsequently incorporated into that. AGAGE and its predecessor programs (ALE and GAGE) have measured the composition of the global atmosphere continuously since 1978 (Prinn et al. 2000). Currently, CMN is an AGAGE-affiliated station: although the instrumentation used is not the same as used at the other AGAGE stations, measurements are linked to the AGAGE calibration scale, and the same calibration and quality assurance protocol as in the AGAGE stations is used.

Radiatively active non-CO_2 gases can be divided into two broad categories, on the base of their capability of destroying the stratospheric ozone: (1) ozone depleting substances (ODSs) regulated under the UNEP Montreal Protocol (MP) and (2) radiatively active gases included in the UNFCCC Kyoto Protocol.

1.1 *Radiatively Active Ozone Depleting Substances*

This definition identifies those halocarbons (i.e., compounds that contain atoms of carbon and halogen atoms) able to destroy the stratospheric ozone layer. These compounds, whose first examples have been developed around 1930 as a safe alternative to toxic and flammable refrigerants, have soon gained a widespread diffusion, and their use has been extended to other applications: aerosol propellants, foam-blowing agents, fire extinguishers, and solvents. These long-lived chemicals are inert in the troposphere, but they undergo photolytic decomposition by ultraviolet (UV) radiation in the stratosphere, becoming a major source of inorganic chlorine and bromine (Molina and Rowland 1974). Some of the released halogen atoms can become active in destroying the stratospheric ozone at polar latitudes (Farman et al. 1985), resulting in more harmful UV-B radiation reaching the Earth's surface and causing biological damage in plants, animals, and humans. The recognition of this phenomenon led in 1987 to the signature of a global environmental treaty, the Montreal Protocol (MP) to Reduce Substances that Deplete the Ozone Layer (UNEP 1987) that had a provision to progressively reduce production levels of the main ozone depleting substances (ODSs): chlorofluorocarbons (CFCs) and halons. Subsequent amendments to the MP called for an accelerated phaseout schedule and extended the number of controlled substances, also including the hydrochlorofluorocarbons (HCFCs), i.e., the chemicals developed by the industry to substitute CFCs. The HCFCs also include hydrogen atoms that can react with tropospheric hydroxyl (OH), resulting in a shorter atmospheric lifetime. However, they retain the capability of releasing chlorine atoms in the stratosphere, even if to a lesser extent with respect to their parent compounds CFCs.

The MP represents a good example of how international agreements can be effective in tackling environmental issues of global significance. In fact, as a consequence of the implementation of the Protocol and its amendments, tropospheric abundances and emissions of most ODSs started to decrease in the mid-1990s, with a resulting decline in global tropospheric chlorine and bromine. In addition, the MP has been highly effective in protecting climate. Indeed, halocarbons are also powerful greenhouse gases characterized by very high GWP values: restricting the use of halocarbons has helped to reduce global warming with the result that the benefit to climate achieved by the Montreal Protocol alone greatly exceeds the initial target of the UNFCCC Kyoto Protocol (Velders et al. 2007, 2012, Montzka et al. 2011).

1.2 Radiatively Active Kyoto Gases

The Kyoto Protocol (UNFCCC 1997) is an international agreement linked to the UNFCCC, which commits its parties by setting internationally binding emission reduction targets for six direct GHGs included in the so-called Kyoto basket: CO_2, CH_4, N_2O, the fluorinated hydrocarbons (hydrofluorocarbons, HFCs, and perfluorocarbons, PFCs), and SF_6. Each gas is weighted by its 100-year GWP and aggregated to give total GHG emissions in CO_2 equivalents. The Protocol was adopted in Kyoto on December 1997 and entered into force on February 2005. Its first commitment period started in 2008 and ended in 2012. During the first commitment period, 37 industrialized countries and the European Community committed to reduce GHG emissions to an average of 5% against 1990 levels. In Doha, in December 2012, an amendment was adopted who agreed to take on commitments in a second period from 1 January 2013 to 31 December 2020. During the second commitment period, parties agreed to reduce GHG emissions by at least 18% below 1990 levels in 8-year period from 2013 to 2020. A further step was made in November 2016, when the Paris Agreement entered into force. The Paris Agreement for the first time brought all nations into a common cause to undertake ambitious efforts to combat climate change and adapt to its effects, with enhanced support to assist developing countries to do so. The Paris Agreement's aim is to strengthen the global response to the threat of climate change by keeping a global temperature rise in this century well below 2 °C above preindustrial levels and to pursue efforts to limit the temperature increase even further to 1.5 °C.

Hydrofluorocarbons (HFCs) The MP has been certainly very successful in phasing out ODSs; however, it led to a shift toward HFCs that, because of their short lifetime and lack of ozone depleting Cl and Br atoms, constitute the best alternative to chlorine- and bromine-containing halocarbons. Like the ODSs that they replaced, HFCs are potent GHGs. For this reason, HFCs have been targeted for emission reductions in the UNFCCC Kyoto Protocol. The increased demand for refrigeration and air conditioning, particularly in developing countries, could lead to a rise in HFC emissions offsetting much of the climate benefit achieved by phasing out the ODSs under the MP. It has been estimated (Montzka et al. 2011; Velders et al. 2012, 2015) that due to the high GWP values of the HFCs, without direct regulation, their atmospheric increase could lead to 14–27% of the increase in CO_2 radiative forcing under the range of IPCC business-as-usual scenarios from 2010 to 2050. Such direct regulation can be achieved only through the mechanisms that have been put in place within the MP. A unique feature of the MP is an adjustment provision that enables the parties to the Protocol to respond quickly to new scientific information, in order to accelerate the reductions required on chemicals already covered by the Protocol. These adjustments are then automatically applicable to all countries that ratified the Protocol. Developing countries are given more time to comply with the phaseout decisions, and also they receive funding from the Multilateral Fund to facilitate compliance with the Protocol's provisions.

The phasedown of HFCs under the MP has been under negotiation by the Parties since 2009 (UNEP 2015a, b, c, d); finally, during the 28th meeting of the parties of the MP (MOP28) in October 2016, 197 countries adopted an amendment to phasedown HFCs. Parties to the MP committed to cut the production and consumption of HFCs by more than 80% over the next 30 years. Developed countries will reduce HFC consumption beginning in 2019, most developing countries will freeze consumption in 2024, and small group of the world's hottest countries (India, Pakistan, Iran, Saudi Arabia, and Kuwait) will freeze HFC use by 2028. In this way up to 0.5 °C warming will be avoided by the end of the century while continuing to protect the ozone layer.

Methane and Nitrous Oxide Differently than the fluorinated gases, which are exclusively anthropogenic, methane (CH_4) and nitrous oxide (N_2O) are two powerful GHGs that also have natural origin.

CH_4, with a 100-year GWP value of 28 (IPCC 2013) and an average global concentration of 1808 ppb in 2012 (Carpenter et al. 2014), is the second most important GHG. Wetlands are its main natural source, while anthropogenic sources include agriculture (e.g., rice production, ruminants), landfills, biomass burning, and the extraction and processing of fossil fuels (Carpenter et al. 2014). CH_4 is not only a powerful GHG but also can indirectly affect climate being a precursor of the radiatively active tropospheric ozone and being able to react with OH radicals. In this way it also influences the lifetimes of the fluorinated gases that have in OH their major sink. Anthropogenic CH_4 emissions have increased significantly since preindustrial times, due to a grown use of fossil fuels, but also due to the increase in ruminants, landfills, and rice fields corresponding to the increase in human population (Ghosh et al. 2015). However, during the 1980s and 1990s, a halt in such emissions was described in several studies (Dlugokencky et al. 1998; Bousquet et al. 2006). After this "pause," CH_4 started to increase again in 2007 (Rigby et al. 2008; Dlugokencky et al. 2009), likely because of increased emissions from natural wetlands and fossil fuels, although their relative contributions remain uncertain (Kirschke et al. 2013). Due to the importance of constraining CH_4 emissions, in the past years, several studies have been performed, and modeling approaches developed (e.g., Bousquet et al. 2006; Bruhwiler et al. 2014; Houweling et al. 2014; Fraser et al. 2013; Meirink et al. 2008).

N_2O has a 100-year GWP value of 265 (IPCC 2013) and a global concentration of 325 ppb in 2012 (Carpenter et al. 2014). The increase of N_2O concentrations from the 271 ppb of the preindustrial to the current ones makes this long-living gas the third most important GHG (IPCC 2013). Beside the natural terrestrial, marine, and atmospheric sources, anthropogenic sources include agriculture, biomass burning, and industry and indirect emissions from reactive nitrogen leaching, runoff, and atmospheric deposition (Carpenter et al. 2014). The use of nitrogen-based fertilizer has been estimated to be the main responsible for the observed atmospheric increase.

In the following paragraphs, the main scientific results obtained based on 15 years of atmospheric measurements of non-CO_2 GHGs at CMN are described. These results have been achieved combining atmospheric data with different modeling approaches. Data from CMN have been employed, mainly in combination with data from other AGAGE stations, to conduct global to regional studies aimed at understanding the budget of a number of radiatively active gases, thus providing scientific policy-relevant decision support information to enhance society's ability to plan and respond.

2 Experimental and Methodologies

2.1 High Frequency Analysis

Non-CO_2 GHGs at CMN are measured via gas chromatographic (GC) methods, using two different instrumentations: a gas chromatograph–mass spectrometer (GC-MS) for the analysis of halogenated gases and a multidetector GC for the analysis of CH_4 and N_2O.

Halogenated Gases Challenges to face in the analyses of this large group of compounds are essentially the wide range of volatilities and atmospheric abundances as well as very low concentration levels, in the range of the part per trillion (ppt, i.e., 10^{-12}). For these reasons, high-precision in situ air analysis requires significant pre-concentration of large sample volumes at sub-ambient temperature and, preferably, without using cryogenic liquids (liquid N_2 or Ar) which are difficult to handle at a remote and often unattended station. As an alternative to cryogenic fluids, an enrichment device operated with a Peltier-cooled pre-concentration technique can be used for adsorbing/desorbing the targeted compounds on a focusing trap filled with a correct combination of adsorbing materials. High-resolution GC separation coupled with the MS detection is required (to extend the best sensitivity and linearity) to all the compounds to be analyzed. Several efforts have focused on achieving this goal, and various adsorbents and chromatographic setting have been tested (see, e.g., Maione et al. (2004)). The instrumentation that is currently in use at CMN is an Agilent 6850–5975 GC–MS equipped with the UNITY2-Air Server2 autosampling/pre-concentration device (Markes International). The pre-concentration of the air samples is done on a focusing trap filled with four different adsorbing materials: Carbograph 2TD, Carbograph 1TD, Carboxen 1000, and Carbosieve SIII. In the sampling/absorption phase, one liter of ambient air is passed through the trap kept at a trapping temperature of −30 °C by means of a three-stage Peltier cell. The trap is then desorbed by heating the trap at 60 °C s^{-1} up to 300 °C and held at this temperature for 2 min. The separation is performed at constant flow on a J&W GS-GasPro capillary column, 30 m × 0.32 mm I.D., at 1.6 ml min^{-1}, i.e., a gas solid chromatographic column with cyclodextrin as stationary phase, working in adsorption gas chromatography (O'Doherty et al. 1999). This column shows high retention and separation factor characteristics, being able to strongly retain the more volatile compounds. The elution order is governed by a combination of factors, i.e., boiling point, hydrogen bonding, and number of chlorine atoms. In addition, this column can strongly absorb molecules forming hydrogen bonds, like those compounds with adjacent hydrogen and fluorine atoms, which is the case for many of the halocarbons. The temperature ramp used is 10 min at 49 °C, then 10 °C min^{-1}, up to 250 °C for a total run time of about 40 min. The MS detector is operating in selective ion mode (SIM); since many compounds have similar electron impact (EI) fragmentation paths, identification and quantification are based on two mass-over-charge ratios (m/z) per compound (target ion and at least one qualifier), in order to improve the diagnostic for determining chromatographic co-elutions. The ambient air sample is analyzed every 2 h and two working standard runs bracket every air sample run, in order to detect and correct for short-term instrumental drift. Measurement repeatability, evaluated from the repeated working standard measurements, goes from 0.3% to 5% for the higher and lower concentration compounds, respectively. Limits of detection, LODs (S/N > 3), are in the range of 0.06–0.2 ppt and have been indirectly estimated on the chromatogram of the working standard run with typical background concentrations for almost all the compounds. The working standards are prepared pumping ambient air at the site during relatively clean background conditions into a 35-l electro-polished stainless steel canister (Essex Cryogenics,

Missouri, USA) using a modified oil-free compressor (SA-3 Rix California, USA) up to ~60 bar. In order to improve the stability of the compounds, the canisters are humidified during the pumping process with purified water (Miller et al. 2008). This ensures a close similarity in composition between the ambient air samples and the reference standard, in order to minimize analytical artifacts and nonlinearity of the method. The working standards are regularly calibrated (almost weekly) against a tertiary standard prepared with the same procedure at the baseline station of MHD and calibrated on the AGAGE scales SIO-05, SIO-07, and SIO-014 developed and maintained at Scripps Institute of Oceanography, UC San Diego (CA, USA), where primary gravimetric standards are prepared. In this way, the contribution of the scale transfer (calibration) uncertainty to the total measurement uncertainty is minimized among the various stations in the network. The system is fully automated and operated via the Linux-based chromatography software GCWerks (gcwerks.com) developed within the AGAGE program at SIO. This methodology allows the simultaneous determination of about 60 volatile organic compounds, of which 30 are halogenated species.

Methane and Nitrous Oxide The analytical system is composed by an Agilent 6890 N gas chromatograph equipped with a flame ionization detector (FID), a carbon monoxide (CO) converter ("methanizer"), and an electron capture detector (micro ECD). Each detection channel consists of a 10 ml sampling loop, a set of pre-column and analytical column mounted on a ten-port two-position switching valve, connected in the backflash configuration to avoid contaminants running through the main columns after the sample injection into the pre-column. The loops are flushed for 3 min with 40 ml min^{-1} before switching the gas selector valve to an off position, waiting 10 s for equilibration. Then, the sampled volume is flushed with the carrier gases into the columns. The separation takes place isothermically at 60 °C on a Unibeads 1S (60/80 mesh, 3 ft × 1/8″) pre-column and a Molecular Sieve 5 Å (60/80 mesh, 4 ft × 1/8″OD) analytical column for the FID channel and on two HayeSep Q (80/100 mesh, 6 ft × 3/16″) columns for the ECD channel, respectively. The ECD channel is used for the analysis of N_2O. CH_4 and CO are analyzed by GC-FID: after the elution of CH_4, the column effluent is diverted by a four-way switching valve to the methanizer: a nickel catalyst heated to 375 °C is continuously flushed with H_2 to reduce the CO to CH_4, subsequently analyzed by the FID.

A ten-port gas selector valve allows to alternately measure ambient air and working standard. The ambient air is supplied to the instrument through stainless steel tubings (1/8″ OD) by means of a KNF Neuberger pump (N86KTE) with a flow rate of approximately 200 ml min^{-1}. A pressure relief valve releases the excess of air maintaining the internal pressure at 0.5–0.8 bar; the flow through the sampling loops is controlled downstream to 40 ml min^{-1} by means of a mass flow controller (MKS). The loops are mounted in an isolated housing on top of the GC oven to reduce temperature fluctuations.

Compressed air (dry) is regularly used as a working standard that is calibrated against a set of NOAA/CMDL standards. The ambient air and the working standards are passed through a Nafion dryer tube mounted between the gas selector valve and the sample loops. Dry zero air for the Nafion counterflow is taken from the zero air generator output. One measurement sequence takes 15 min. The system is fully automated via the GC Manager software adopted from the AGAGE program.

2.2 *Baseline and Trend Evaluation*

The analysis of the time series obtained through the methods described in Sect. 2.1 is frequently based on the identification of a baseline.

Baseline concentration for a given compound is defined as that which is not significantly affected by non-well-mixed contributions. Therefore, baseline concentration evaluated at CMN should not have been influenced by local or regional emissions, in order to be representative of the European background for the given compound.

We implemented a statistical approach to separate the non-well-mixed contributions from those belonging to the background. Considering that a well-mixed pollutant (even in a non-stationary state) follows a Gaussian distribution, we can assume that the overall error, given by the instrumental error and by the natural background variability, follows a Gaussian distribution. A gamma distribution can be used to simulate the probability density function (PDF) of the non-well-mixed contributions (i.e., pollution events). The proposed approach is based on two steps. The concentration time series are first detrended on an appropriate time interval. After that, the observed PDF (i.e., detrended data) is fitted by the sum of a Gaussian and a gamma PDF, applying a maximum likelihood estimation function (Statistics Toolbox™). Finally, we assign to the baseline all the mixing ratio values for which the corresponding Gaussian PDF is higher than the Gamma one. A detailed description of the approach is given in Giostra et al. (2011).

Background value monthly means are used to build halocarbons long-term time series. The best fits of the monthly means of baseline data are obtained applying the polynomial in Eq. (1), as in Simmonds et al. (2004):

$$\chi(t) = a + bNP_1\left(\frac{t}{N}-1\right) + \frac{1}{3dN^2P_2}\left(\frac{t}{N}-1\right) + c\cos(2\pi t') + s\sin(2\pi t') \quad (1)$$

where χ are monthly mean baseline mole fractions after pollution events have been removed. The coefficients a, b, and d define the average mole fraction (ppt) over the investigating period, the trend in mole fraction (ppt year^{-1}), and the acceleration in the trend (ppt year^{-2}), respectively. c and s define the annual cycle in dry air mole fractions, and t is the time measured from the beginning of the 2 N-year interval of interest; the zero of t' is always July 1 so that the phases of the seasonal cycles may be easily compared. The normalized root mean square error (NRMSE) of simulated (S) values against the observed (O) ones is calculated as in Eq. (2):

$$\mathrm{NRMSE} = \frac{\sqrt{\frac{\sum_{i=1}^{n}\left(O_i - S_i\right)^2}{n}}}{\bar{O}} \tag{2}$$

2.3 Emission Allocation and Quantification on a European Scale

The estimate of the emission fluxes of radiatively active species on a European scale is one of the most significant applications of the observation activity described above. Most of the regional studies performed on the species measured at CMN are based on similar approaches, whose detailed description is given in the related papers, discussed in Sect. 3. Here we describe the general structure of the most used modeling procedure.

Dispersion Modeling and Bayesian Inversion Allocation and quantification of emissions are performed through a multistep procedure based on (1) a Lagrangian particle dispersion model (LPDM) that calculates backward particle trajectories from a receptor (or more receptors) and produces the emission sensitivity map of the source–receptor relationships (in other word, what fraction of a unitary emission from a given source reaches the receptor). Therefore, a simulated mixing ratio contribution of the cell at the receptor is given: (2) an a priori emission field with assigned spatial uncertainties. Combining the a priori emission field with the particle trajectories, the time series of a concentration first guess at the receptor is obtained; (3) a Bayesian inversion technique in which coupling of the concentration time series measured at the receptor with the first guess time series (and its associated uncertainties) reconstructs the most probable a posteriori emission field.

Backward particle trajectories of the FLEXPART LPDM based on the European Centre for Medium-Range Weather Forecast wind fields are used (Stohl et al. 2005).

The estimate of the a priori emission field and of the related uncertainty represents a delicate step in the procedure, since it will determine the reliability of the a posteriori field produced by the Bayesian inversion. The a priori emission fields are built using different datasets according to information that is available for each specific compound. Bottom-up inventories as well as outputs of global scale model can be used, and they have been spatially allocated based on the specific use of the compound (e.g., on the base of population distribution or industry distribution).

Finally, the a posteriori emission flux is evaluated applying the Bayesian inversion method proposed by Seibert (2000, 2001) modified by Eckhardt et al. (2008) and improved by Stohl et al. (2009). It should be highlighted that the number and the location of the receptors strongly affect the reliability of the a posteriori emission flux on the entire European domain.

3 Results

3.1 *Ozone Depleting and Radiatively Active Gases*

Most of the ODSs included in the MP are monitored at CMN, and their list, characteristics, and phaseout schedule are given in Table 1.

In a study aimed at assessing the effectiveness of the MP from a regional perspective, we calculated trends and annual growth rates up to December 2012, comparing the results obtained with those from other international observation programs (Maione et al. 2013). In Fig. 1 we report the ODS time series and trends updated up to June 2016. Measurements have been carried out following the methodology described in Sect. 2.1. The polynomial parameters for each ODS, obtained through the procedure described in Sect. 2.2, are summarized in Tables 2a, 2b, along with the normalized root mean square error (NRMSE).

CFCs Four CFCs are monitored at CMN: the average atmospheric mixing ratios updated to 2016 are 529.1 ppt for CFC-12, 245.7 ppt for CFC-11, 16.5 for CFC-114,

Table 1 Ozone depleting and radiatively active substances measured at CMN and their ozone depleting potential (ODP), global warming potential (GWP) at 20 and 100 years, lifetime, main industrial, and/or commercial uses

Compound	ODP	GWP 20 years	GWP 100 years	Lifetime (year)	Main uses
CFC-11	1	6900	4660	45	Aerosol propellants; open-cell foam blowing; closed-cell foam-blowing agent
CFC-12	1	10,800	10,200	100	Refrigerant; aerosol propellants; open-cell foam blowing
CFC-114	1	7710	8590	190	Refrigerant; propellants in medical aerosols
CFC-115	0.6	5860	7670	1020	Refrigerant
Halon-1211	3	4590	1750	16	Fire extinguisher
Halon-1301	10	7800	6290	65	Fire extinguisher
HCFC-22	0.055	5280	1760	11.9	Refrigerant, air conditioning, extruded polystyrene foam applications
HCFC-142b	0.065	5020	1980	17.2	Blowing agent in extruded polystyrene board stock
HCFC-124	0.022	1870	527	5.9	Refrigerant
Methyl chloroform	0.1	578	160	5	Industrial solvent
Carbon tetrachloride	1.1	3480	1730	26	Raw material in industrial uses
Methyl bromide	0.6	9	2	0.8	Fumigant in agriculture; quarantine and pre-shipment applications

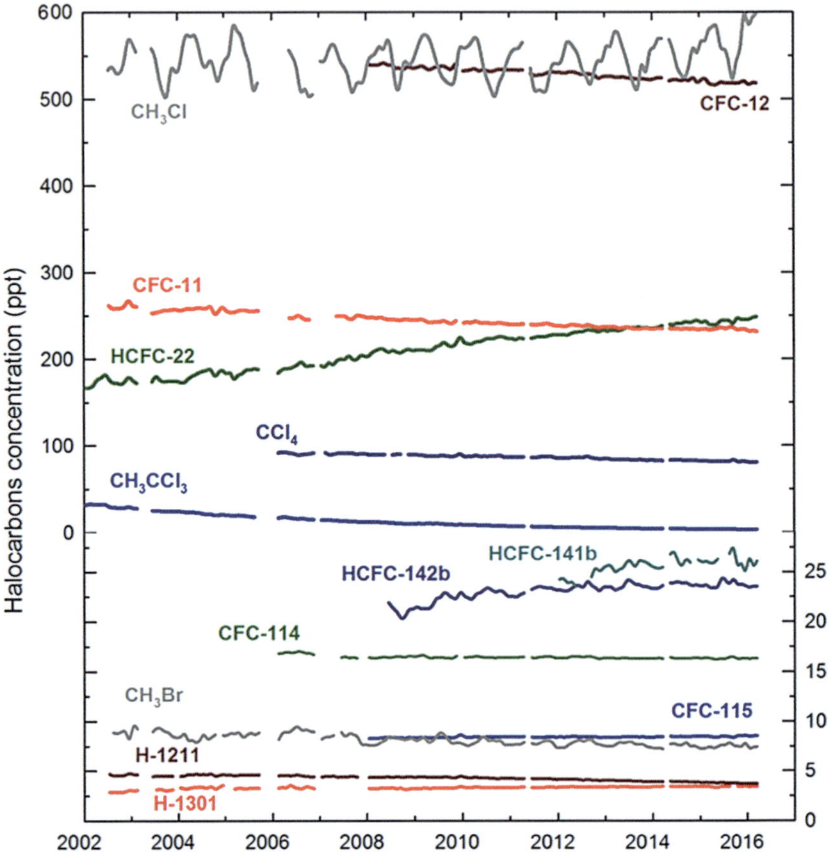

Fig. 1 Time series of the ozone depleting radiatively active gases measured at CMN

and 8.4 ppt of CFC-115. Due to their past widespread use and long atmospheric lifetimes, CFCs are still the most abundant ODSs in the atmosphere. Effects of their complete phaseout under MP in 1996 in non-Article 5 (non-A5) and in 2010 in Article 5 (A5) countries are evident in the decline of their background atmospheric mixing ratios of −0.9, −0.5, −0.3% year^{-1}, for CFC-11, CFC-12, and CFC-114, respectively, over the periods for which measurements at CMN are available (see Tables 2a, 2b). For CFC-115, the longest living among the CFCs considered, a still positive trend of 0.2%, is observed but with no acceleration. The measured average mixing ratios are in agreement with the global ones (Carpenter et al. 2014) confirming the efficient mixing of these compounds in the atmosphere at a global scale, deriving from their prolonged atmospheric lifetimes and by the switching off of their sources. The progressive decrease in the last 6 years of data above the baseline is a clear indication of the nearly complete turning off of emission sources in the station domain. Nevertheless, CFC emissions still occur as shown in a study by Keller et al. (2012) based on measurements at four European sites (CMN included)

Table 2a Polynomial coefficients

	a, average ppt	b, trend ppt $year^{-1}$	trend %$year^{-1}$	c, trend acceleration ppt $year^{-2}$
CFC-11	245.7 ± 0.3	−2.17 ± 0.08	−0.90 ± 0.03	0.16 ± 0.04
CFC-12	529.1 ± 0.3	−2.8 ± 0.1	−0.50 ± 0.03	−0.14 ± 0.14
CFC-114	16.5 ± 0.02	−0.04 ± 0.01	−0.30 ± 0.04	0.02 ± 0.01
CFC-115	8.40 ± 0.01	0.010 ± 0.004	0.20 ± 0.05	0 ± 0
HCFC-22	209.1 ± 0.5	5.8 ± 0.13	2.8 ± 0.1	−0.11 ± 0.07
HCFC-141b	2.37 ± 0.01	0.20 ± 0.01	8.4 ± 0.5	−0.01 ± 0.02
HCFC-142b	22.95 ± 0.06	0.36 ± 0.03	1.5 ± 0.1	−0.15 ± 0.03
H-1211	4.30 ± 0.01	−0.06 ± 0.002	−1.50 ± 0.05	0 ± 0
H-1301	3.30 ± 0.02	0.020 ± 0.004	0.70 ± 0.12	0.003 ± 0.003
CCl_4	86.6 ± 0.1	−1.06 ± 0.04	−1.20 ± 0.05	−0.06 ± 0.03
CH_3Cl	546.2 ± 1.7	0.4 ± 0.4	0.10 ± 0.1	0.9 ± 0.2
CH_3CCl_3	12.6 ± 0.1	−1.99 ± 0.02	−15.8 ± 0.2	0.32 ± 0.01
CH_3Br	8.2 ± 0.1	−0.12 ± 0.01	−1.5 ± 0.2	−0.001 ± 0.009

For explanation of terms, see text

Table 2b Polynomial coefficients

	cos, seasonal cycle	sen, seasonal cycle	NRMSE %	Measurement period
CFC-11	-0.24 ± 0.42	−0.18 ± 0.43	0.7	2002–2016
CFC-12	-0.45 ± 0.47	−0.13 ± 0.48	0.3	2008–2016
CFC-114	0.003 ± 0.028	−0.005 ± 0.028	0.6	2006–2016
CFC-115	0 ± 0	0 ± 0	0.5	2008–2016
HCFC-22	0.95 ± 0.74	0.50 ± 0.76	1.6	2002–2016
HCFC-141b	0.02 ± 0.02	−0.01 ± 0.02	1.7	2012–2016
HCFC-142b	0.28 ± 0.09	0.06 ± 0.09	1.5	2008–2016
H-1211	0 ± 0	0 ± 0	1.1	2002–2016
H-1301	0 ± 0	0 ± 0	2.6	2002–2016
CCl_4	0.04 ± 0.18	0.21 ± 0.18	0.7	2006–2016
CH_3Cl	−8.8 ± 2.3	−23.5 ± 2.4	2.2	2002–2016
CH_3CCl_3	−0.1 ± 0.1	−0.3 ± 0.1	4.0	2002–2016
CH_3Br	0.20 ± 0.08	0.07 ± 0.08	4.0	2002–2016

For explanation of terms, see text

combined with the LPDM FLEXPART and a Bayesian inversion technique. Total emissions for a study area including most European countries, Scandinavian region excluded, have been estimated to be in 2009 4.2 (2.9–5.4) Gg $year^{-1}$ and 2.2 (1.0–3.4) Gg $year^{-1}$ for CFC-11 and CFC-12, respectively (Keller et al. 2012).

Halons The presence of bromine atoms makes halons particularly effective in destroying the stratospheric ozone. Their complete phaseout was scheduled in 1994 and 2010 in non A-5 and A-5 countries, respectively. Since their use is restricted almost exclusively as fire extinguishing agents, their atmospheric mixing ratios are in the range of a few ppt. However, due to this specific use, the estimation of the amounts of halons stocked in banks is not easy, and it is likely to be still significant (UNEP 2007) as also shown by the pollution episodes still recorded at CMN. Updated H-1211 and H-1301 average mixing ratios measured at CMN during 2002–2016 are 4.3 and 3.3 ppt, respectively. In agreement with global average concentrations, H-1211 seems to have started a steady decline around 2005, and currently its atmospheric trend is −1.5% $year^{-1}$. H-1301 shows an overall trend of 0.7% $year^{-1}$ but with no acceleration since 2010, suggesting that the peak in the atmospheric mixing ratio has probably been reached. Although halons have been banned in Europe since 1994, still occurring elevations above the baselines at CMN suggest the presence of some fresh emissions within the station domain, even if the pollution episodes have declined in the last few years.

Keller et al. (2012) have estimated in 2009 H-1211 emissions ranging from 26 (15–38) Mg $year^{-1}$ in central west to 54 (40–69) Mg $year^{-1}$ in northeast Europe.

Halon data from CMN have been recently incorporated into a global study (Vollmer et al. 2016) where measurements carried out at the AGAGE and NOAA stations, as well as in archived air in canisters and in polar firn samples, have contributed in reconstructing atmospheric histories and global emissions of these compounds. Continuing emissions of H-1211 and H-1301 near in most AGAGE sites have been shown. The authors found that for H-1211, after a peak of 11 Gg $year^{-1}$ occurred in the late 1990s, emissions declined to 3.9 Gg $year^{-1}$ in the period 2013–2015. For H-1301 a peak in emissions of 5.4 Gg $year^{-1}$ was reached in the late 1980s, and currently (2013–2015) emissions have been estimated in 1.6 Gg $year^{-1}$. Even if summed halon emissions have decreased substantially, nevertheless since 2000, they have accounted for ca. 30% of the emissions of all major anthropogenic ozone depletion substances, when weighted by ozone depletion potentials.

HCFCs The complete phaseout date of these second-generation chlorofluorocarbons has not yet been reached in A-5 parties, and their background concentrations are still increasing. Average concentrations at CMN are 209.1, 25.6, and 23.0 ppt for HCFC-22 (2002–2016), 141b (2012–2016), and 142b (2008–2016) with trends of +2.8, +1.8, and +1.5% ppt y^{-1}, respectively.

CMN data for HCFC-22, the most abundant HCFC in the atmosphere, have contributed to several studies. Saikawa et al. (2012) published a very comprehensive paper on global and regional emission estimates, based on HCFC-22 atmospheric data from the AGAGE and NOAA networks combined with a three-dimensional chemical transport model and a Bayesian inverse method. The inversion indicated that global HCFC-22 emissions had an increasing trend over the study period (1995 to 2009), with a further surge between 2005 and 2009 as a consequence of increasing emissions from developing countries in Asia. A study focused on the European scale (Graziosi et al. 2015) has been based on observations at CMN and MHD, the

FLEXPART dispersion model, and a Bayesian inversion technique. European fluxes of HCFC-22 have been estimated over an 11-year period from January 2002 to December 2012; during which, the accelerated restrictions on HCFCs production and consumption have entered into force under the MP. According to this study, the maximum of emissions from the European domain was in 2003 (38.2 ± 4.7 Gg $year^{-1}$) and the minimum in 2012 (12.1 ± 2.0 Gg $year^{-1}$); emissions continuously decreased between these years, except for secondary maxima in the 2008 and 2010. The largest fraction of European emissions was attributed to two regions, France and a large macro-area in Eastern Europe, with highest per-capita emissions in France and the United Kingdom. Despite such a decrease in regional emissions, European background values of HCFC-22 over 2002–2012 are still increasing as a consequence of global emissions, in part from developing countries. However, a decrease in the atmospheric growth rate of HCFC-22 has been recorded since then, suggesting that it is likely that global emissions are not growing as rapidly as until 2008. In addition, three monthly inversions allowed the identification of a seasonal cycle in HCFC-22 emissions in Southern Europe: in the southernmost countries adjacent to the Mediterranean basin, the emissions in summer were ca. 25% higher than those in the colder months, likely because of a more intensive use of refrigeration and air-conditioning systems during the warm season. The European emissions reported in this study were in a very good agreement with those reported in some previous ones, also based on CMN data, i.e., European estimates for 2005 and 2006 published by Stohl et al. (2009) and those for 2009 by Keller et al. (2012), but considerably larger than estimates reported by Saikawa et al. (2012) for Europe.

Methyl Chloroform Methyl chloroform (MCF) is a man-made chlorinated solvent controlled under the MP that called for its phaseout in 1996 in developed countries and 2015 in developing countries. Updated average atmospheric mixing ratios at CMN are 12.6 ppt, with a decreasing trend since 2002 of 15.8% $year^{-1}$. With OH being the main removal mechanism for MCF in the troposphere, long-term global measurements of MCF, combined with emission estimates, have been used to infer both hemispheric and global average mole fractions and trends, as well as variability of the OH radical on annual timescales (see, e.g., Montzka et al. 2000; Prinn et al. 2001, 2005) that would be otherwise difficult because no globally representative measurements of OH exist. Due to its capability of removing many radiatively important trace gases and atmospheric pollutants, OH play a key role in atmospheric chemistry. Because of the crucial role of MCF emissions in the evaluation of hemispheric and global mean OH concentrations, in the past years, a scientific debate took place on the actual extent of emissions, in particular on a European scale. European emission fluxes estimates went from 0.3 to 3.4 Gg $year^{-1}$ in 2000–2004 (Reimann et al. 2005) to more than 20 Gg in 2000 (Krol et al. 2003). The persisting occurrence of non-negligible mixing ratio enhancements of MFC at the European sites of CMN and JFJ suggested unexpectedly high on-going emissions somewhere in Europe, despite MCF phaseout date of 1996. In order to identify the source region and estimate the magnitude of such emissions, a Bayesian inversion method (Stohl et al. 2009) as well as a point source analysis (Keller et al. 2011) have been used in

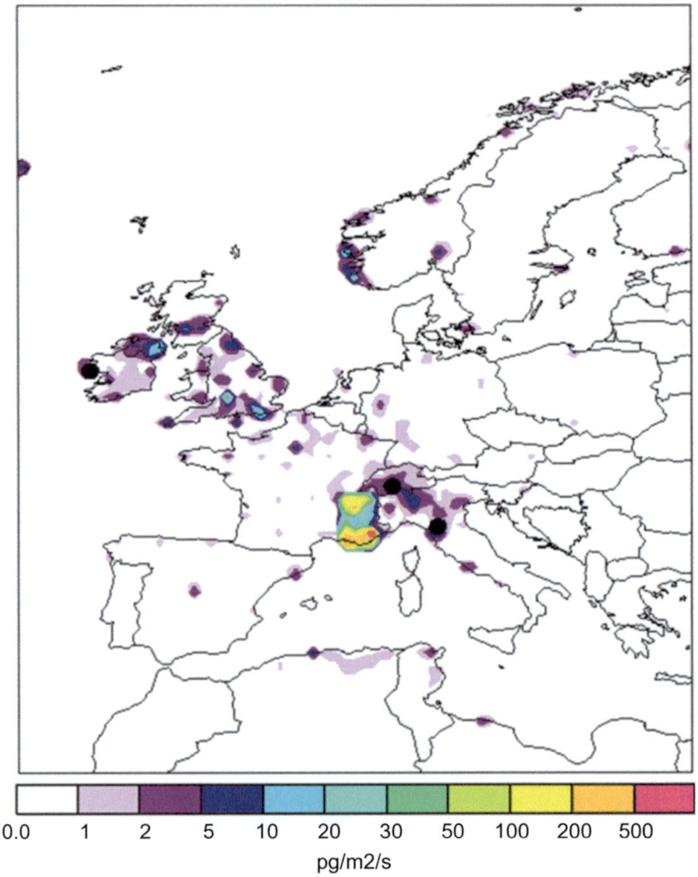

Fig. 2 A posteriori methyl chloroform emissions in pg m^{-2} s^{-1} based on measurements at three European sites. Reference period: January 2008–December 2008. *Black dots* indicate the locations of the measurements sites (Adapted from Maione et al. 2013)

combination with observations at CMN and other two European sites (JFJ and MHD) (Maione et al. 2014). This study revealed that over the study period (2002–2012), the strongest sources of MCF in Europe are located in southeastern France, where landfill and waste disposal and halogen chemical plants are present, as shown in the emission map in Fig. 2. This very well-defined area (30% of uncertainty) accounted for a large fraction of global emissions during this period, ranging from 2.6% in 2003 to 10.3% in 2009 of the MCF global emission estimates.

Carbon Tetrachloride CCl_4 is a near exclusively anthropogenic compound used as solvent, feedstock for chlorinated chemicals production, fire extinguisher, fumigant, and rodenticide. It is a very powerful GHG (GWP 1730) and ODS (ODP 0.72). CCl_4 is measured at CMN since 2006, and average background concentration is 86.7 ppt with an atmospheric trend of −1.2% over the measurement period. Currently, emissive uses of CCl_4 are banned under the MP, but production and use are still allowed for feedstock use. Chemical feedstock should be converted into new chemicals, effectively destroying the feedstock, but fugitive emissions are possible. Other possible sources for CCl_4 beside fugitive emissions from the industry sector (Simmonds et al. 1998; Fraser et al. 2014) are generation during bleaching (Odabasi et al. 2014) or emissions from old landfill (Fraser et al. 2014). The discrepancy between global emissions as derived by bottom-up methods and those based on atmospheric observations (the latest one order of magnitude larger with respect to the previous) suggests on-going emissions. This is reinforced by the persistence of an interhemispheric gradient of about 1.3 ppt since 2006 (Carpenter et al. 2014). In order to better constrain the atmospheric budget of CCl_4, several studies based on a combination of atmospheric observations and inverse modeling have been conducted in recent years in various regions of the world (Xiao et al. 2010a; Vollmer et al. 2009; Miller et al. 2012; Fraser et al. 2014; Hu et al. 2016; SPARC 2016). In a study focused on the European scale and based on long-term high-frequency observations at CMN, JFJ, and MHD combined with a Bayesian inversion methodology, average European emissions for 2006–2014 were evaluated to be 2.2 (± 0.8) Gg year^{-1}, with an average decreasing trend of 6.9% per year (Graziosi et al. 2016). This analysis identified France as the main source of emissions over the whole study period, with an average contribution to total European emissions of approximately 26%. The inversion was also able to allow the localization of emission "hotspots" in the domain, with major source areas in Southern France, Central England (UK), and BE-NE-LUX (Belgium, The Netherlands, Luxembourg), where most of industrial scale productions of basic organic chemicals are located. According to this study, European emissions correspond on average to 4.0% of global emissions for 2006–2012.

Methyl Chloride CH_3Cl is a primarily natural compound whose production and consumption are not controlled by the MP. However, CH_3Cl, the largest natural source of stratospheric chlorine, is expected to play an increasing role in determining future levels of stratospheric chlorine as the impact of the anthropogenic MP species declines. In addition CH_3Cl is a radiatively active gas with a 20-year. GWP value of 45 (IPCC 2013). CH_3Cl is measured at CMN since 2002, with an average baseline concentration of 546.2 ppt, a negligible trend of 0.1. CH_3Cl measurements at CMN have been included in a global study where the surface fluxes of methyl chloride were estimated using a 3-D global chemical transport model (Xiao et al. 2010b). Authors estimated global total emissions around 4100 Gg year^{-1} with very large emissions of 2200 Gg year^{-1} from tropical plants, which turn out to be the largest single source in the CH_3Cl budget.

Methyl Bromide Methyl bromide is an ozone depleting substance with both natural and anthropogenic sources. In the past, the primary anthropogenic sources of CH_3Br have been from its use as a fumigant in agriculture and non-quarantine pre-shipment (non-QPS) fumigation uses. These sources are subject to phase out under the MP which called for its complete phaseout in 2005 and in 2015 in non A-5 and A-5 countries, respectively. However, several EU Member States asked and obtained exemptions, until March 2010, for critical uses, i.e., cultivation of high value-added agricultural products. Such exemptions largely interested Italy and Spain and, to a smaller extent, other EU countries. Quarantine and pre-shipment (QPS) uses, which are mainly related to pest control during transport and storage, are exempted from the phaseout, and according to UNEP (2014), consumption from QPS uses was approximately stable in the last two decades. Due to the importance of the Mediterranean region as a source area for both natural and anthropogenic CH_3Br, measurements at CMN are of particular interest. CH_3Br measurements at CMN started in 2002: average background concentration is 8.2 ppt and the trend is −1.4% year^{-1}. As a consequence of the MP, a decrease of pollution episodes is observed at CMN in the period 2008–2009, while a renewed increase of such episodes occurred in 2010–2011, i.e., after the complete phaseout and also after the cessation of allowed exemptions (Maione et al. 2013). The increase of fresh emissions can be probably ascribed to massive and unreported use of stocks.

3.2 *Radiatively Active Kyoto Gases*

Effective monitoring of HFC emissions is critical for tracking progress toward the achievement of emission reduction targets within the Kyoto Protocol regulatory frameworks. HFC monitoring activity at CMN has been especially aimed at providing emission estimates in support of the quality assurance of the inventories, whose uncertainty is only loosely constrained among different countries (IPCC 2006). HFCs are considered particularly suitable for inverse modeling studies as they are solely of anthropogenic origin and sufficiently long lived. In Table 3 the list of HFCs monitored at CMN and their characteristics are given.

Table 3 Hydrofluorocarbons measured at CMN and their characteristics

Industrial name	Chemical formula	Lifetime (years)	GWP 20 years	GWP 100 years	Main use
HFC-32	CH_2F_2	5.2	2430	677	Refrigerant
HFC-125	CHF_2CF_3	28.2	6090	3170	Refrigerant
HFC-134a	CH2FCF3	13.4	3710	1300	Refrigerant
HFC-143a	CH3CF3	47.1	6940	4800	Refrigerant
HFC-152a	CH3CHF2	1.5	506	138	Foam blowing
HFC-227ea	CF3CHFCF3	38.9	5360	3350	Fire retardant
HFC-236fa	CF3CH2CF3	242	6940	8060	Fire retardant, refrigerant
HFC-245fa	CHF2CH2CF3	7.7	2920	858	Foam blowing
HFC-365mfc	CH3CF2CH2CF3	8.6	2660	804	Foam blowing

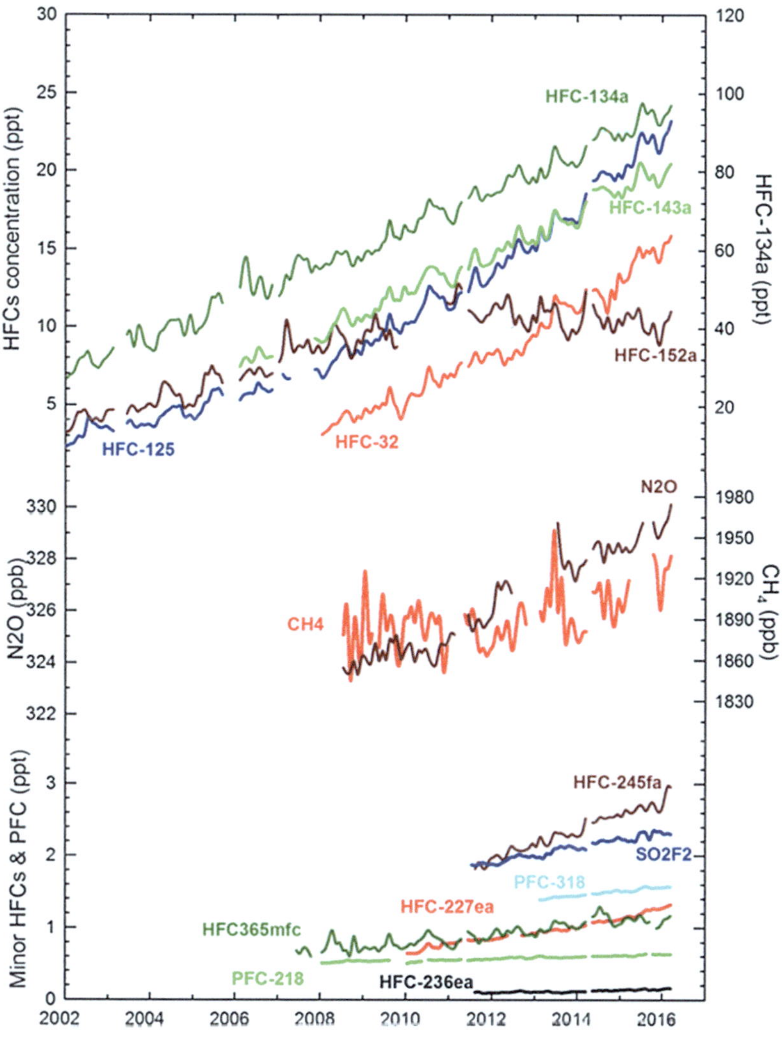

Fig. 3 Time series of the non-CO_2 Kyoto gases measured at CMN

In the following, results obtained based on the observations of Kyoto gases (excluding CO_2) measured at CMN are reported. Their time series and related parameters are given in Fig. 3 and in Tables 4a, 4b, respectively. An increase of all the HFC mixing ratios is observed as a consequence of the phaseout of CFCs and HCFCs under the MP. The results obtained so far for the main HFCs measured at CMN are reported in the following.

HFC-134a HFC-134a is the most abundant HFC in the global atmosphere (Carpenter et al. 2014). Main uses are as replacement refrigerant for CFC-12 in mobile air conditioners, in stationary refrigeration, as well as in foam-blowing

Table 4a Polynomial coefficients for the Kyoto gases monitored at CMN

Compound	a, average ppt	b, trend ppt year^{-1}	trend,%year^{-1}	c, trend acceleration, ppt year^{-2}
HFC-32	8.6 ± 0.1	1.46 ± 0.05	17 ± 0.5	0.14 ± 0.05
HFC-125	10.2 ± 0.1	1.39 ± 0.01	13.6 ± 0.1	0.16 ± 0.01
HFC-134a	61.5 ± 0.3	4.7 ± 0.07	7.6 ± 0.1	0.05 ± 0.04
HFC-143a	13.6 ± 0.1	1.28 ± 0.03	9.5 ± 0.2	0.03 ± 0.02
HFC-152a	8.5 ± 0.1	0.53 ± 0.02	6.3 ± 0.3	−0.12 ± 0.01
HFC-227ea	0.95 ± 0.01	0.09 ± 0.01	10 ± 0.4	0.01 ± 0.01
HFC-236fa	0.18 ± 0.02	0.02 ± 0.01	13.3 ± 3.7	0 ± 0
HFC-245fa	2.26 ± 0.01	0.21 ± 0.01	9.3 ± 0.5	−0.02 ± 0.02
HFC-365mfc	0.89 ± 0.01	0.06 ± 0.01	6.6 ± 0.6	0 ± 0
PCF218	0.572 ± 0.002	0.015 ± 0.001	2.6 ± 0.2	0 ± 0
SO_2F_2	2.065 ± 0.007	0.11 ± 0.01	5.3 ± 0.3	−0.01 ± 0.01
CH_4	1881.5 ± 3.5	9.3 ± 1.5	0.5 ± 0.1	2.2 ± 1.4
N_2O	326.5 ± 0.1	0.8 ± 0.03	0.3 ± 0.01	0.1 ± 0.03

For explanation of terms, see text in Sect. 2.2

Table 4b Polynomial coefficients for the Kyoto gases monitored at CMN

Compound	cos, seasonal cycle	sen, seasonal cycle	NRMSE %	Measurement period
HFC-32	0.24 ± 0.15	−0.11 ± 0.16	6.3	2008–2016
HFC-125	0.48 ± 0.08	−0.02 ± 0.08	4.9	2002–2016
HFC-134a	2.01 ± 0.42	−0.36 ± 0.43	3.8	2002–2016
HFC-143a	0.5 ± 0.1	−0.09 ± 0.1	3.7	2006–2016
HFC-152a	0.06 ± 0.14	−0.65 ± 0.15	7.3	2002–2016
HFC-227ea	0.01 ± 0.01	0.01 ± 0.01	2.6	2010–2016
HFC-236fa	0 ± 0	0 ± 0	4.0	2012–2016
HFC-245fa	0.02 ± 0.02	−0.02 ± 0.02	2.1	2012–2016
HFC-365mfc	0.05 ± 0.02	−0.02 ± 0.02	8.2	2008–2016
PCF218	0 ± 0	0 ± 0	1.8	2008–2016
SO_2F_2	0 ± 0	0 ± 0	1.4	2012–2016
CH_4	−3.6 ± 4.8	7.9 ± 4.9	1	2008–2016
N_2O	−0.1 ± 0.1	0.1 ± 0.1	0.1	2008–2016

For explanation of terms, see text in Sect. 2.2

applications, in aerosol inhalers, and for dry etching. In Europe HFC-134a is mainly (48–59% in 2010) used as refrigerant fluid in Mobile Air Conditioning (MAC) systems. Average HFC-134a background concentration at CMN since 2002 is 61.5 ppt with an increasing trend of 7.6%.

HFC-134a emission estimates also based on observation at CMN have been reported in several studies. Stohl et al. (2009) developed their analytical inversion method to determine the regional and global emissions of halocarbons in 2005 and 2006, exploiting observations from the AGAGE and NIES networks. The method builds on backward simulations with the FLEXPART Lagrangian particle dispersion model; emission information are extracted from the observed concentration increases over a baseline that is itself objectively determined by the inversion algorithm. Resulting global HFC-134a emissions increased from 130 Gg $year^{-1}$ in 2005 to 140 Gg $year^{-1}$ in 2006. In Europe, emissions raised from 24 in 2005 to 27 Gg $year^{-1}$ in 2006, resulting higher than the industry-based a priori emissions. According to this study, in many western European countries (e.g., France, Ireland, Spain, the United Kingdom), emissions were slightly decreasing from 2005 to 2006, while in some southern (e.g., Italy) and eastern (e.g., Poland) European countries, emissions increased. In general a good agreement between UNFCCC reported emissions and the inversion estimates was found. Somewhat different results were obtained by Keller et al. (2012) that estimated for 2009 European emissions (excluding the Scandinavian region) of 22.3 Gg $year^{-1}$. Such emissions in eastern Europe were significantly higher (up to ten times more) than those reported to UNFCCC, while in central west Europe, emissions were less than half of the respective inventories. Also Lunt et al. (2015), using AGAGE and NIES observations and two chemical transport models (Numerical Atmospheric-dispersion Modelling Environment, NAME, and Model for OZone and Related Tracers, MOZART) estimated that in 2007–2012, differently than other HFCs, HFC-134a emissions from Annex I (developed) countries were only 79% (63–95%) of the UNFCCC inventory total. On the same line, Graziosi et al. (2017) found, over the period 2003–2014, European emissions 25% lower than UNFCCC, suggesting that emissions reported by European countries to UNFCCC are likely to be affected by an overestimate of the emission factors. This is especially true for the Scandinavian region, the United Kingdom, and Germany. The only exception is Italy that reports emissions that are two thirds of the estimated ones. This is probably due to the different emission factors used by the Italian authority in charge for the inventory compilation. Differently than UNFCCC, annual emissions derived by the inversion did not show any statistically significant trend. The estimated emissions correspond to the 47.1% of the total European HFC emissions in mass and the 30.3% in CO_2 equivalent (over 2008–2014).

HFC-32, HFC-125, and HFC-143a These three compounds are mostly used in blend for stationary air conditioners and for minor applications (e.g., for fire suppression) as replacement of HCFC-22 and CFC-115 (IPCC/TEAP 2005). Their background concentrations at CMN are 8.6, 10.2, and 13.6 ppt for HFC-32, HFC-125, and HFC 143a, respectively, over the periods reported in Tables 2a, 2b. Reflecting the global situation, these three HFCs show the highest increasing trend of all F-gases (17%, 13.6%, and 9.5%, respectively), showing the importance of these refrigeration blends as replacement of HCFCs and CFCs. According to Graziosi et al. (2017), together their emissions account for more than one third in mass of the most important HFCs emitted in Europe and, due to the high GWP of HFC-143a, for nearly two thirds in CO_2 equivalent over 2008–2014. Their emissions as reported in the same study are in general in line with bottom-up inventories submitted to UNFCCC. Other estimates based on CMN data includes the above-cited regional study by Keller et al. (2012) and the global study by Lunt et al. (2015).

HFC-152a It is used mainly as foam-blowing agent and aerosol propellant (Greally et al. 2007), and its average background concentration at CMN is 8.5 ppt with an increasing trend of 6.3% from 2002 to 2016.

Data from CMN have been included in a number of regional studies estimating European emissions at 4.0 Gg in 2005–2006 (Stohl et al. 2009) and 2.9 Gg in 2009 (Keller et al. 2012). More recently Simmonds et al. (2016) estimated emissions ranging from 4.0 Gg year^{-1} in 2006 to only 2.5 Gg year^{-1} in 2014, and Graziosi et al. (2017) estimated over 2003–2014 average emissions of 4.1 Gg year^{-1}, with a decreasing trend of −4.6% year^{-1}. In comparing all these studies, it should be taken into account that the inversions have been run for different domains. Due to the low GWP of HFC-152a, its averaged relative percent share to overall HFC emissions from Europe over 2008–2014 in Gg-CO_2-eq·year^{-1} is only 0.6% (Graziosi et al. 2017).

HFC-245fa and HFC-365mfc These are replacements for HCFC-141b in foam-blowing applications (Stemmler et al. 2007; Vollmer et al. 2011). These compounds have been introduced into the market more recently with respect to the HFCs previously described, and therefore shorter time series are available. HFC-245a is measured at CMN since 2011 with an average concentration of 2.3 ppt, an increasing trend of 9.3% year^{-1} and estimated European emissions during 2008–2014 of 0.74 Gg year^{-1}, in fair agreement with what is reported to UNFCCC (Graziosi et al. 2017). HFC-365mfc is measured since 2007; its average concentration is 0.9 ppt, the trend 6.6% year^{-1}, and average emissions over 2005–2014 of 1.2 Gg year^{-1}, a value that is three times than that given by the UNFCCC (Graziosi et al. 2017).

HFC-227ea is used in fire suppression, metered-dose inhalers, refrigeration, and foam blowing. Measurement at CMN started in 2010, giving an average concentration of 1 ppt and a trend of 10% year^{-1}. 2008 to 2014 European emissions have been estimated by Graziosi et al. (2017) at an average value of 0.41 Gg year^{-1}, with a positive trend of 4.4% year^{-1}, results that are in a very good agreement with the UNFCCC annual emission values.

HFC-236fa is a fire retardant and a coolant in specialized applications, measured at CMN since 2011. Average concentrations are 0.2 ppt, increasing on average of 13.3% year^{-1}. Average European estimated emissions over 2008–2014 are 0.02 Gg year^{-1}, with a trend of 3.6% year^{-1}, exceeding UNFCCC country report by about 50%. This suggests a deficit in the UNFCCC inventories, with very few countries reporting their emissions.

The geographic distribution of per-capita emissions (in kg-CO_2-eq·year^{-1} ·inhabitants^{-1}) of nine HFCs during 2008–2014 from 12 macro-areas in the European geographic domain is shown in Fig. 4. Column "A" refers to the top-down inversion estimates as derived by Graziosi et al. (2017) and column "B" to UNFCCC bottom-up country reports.

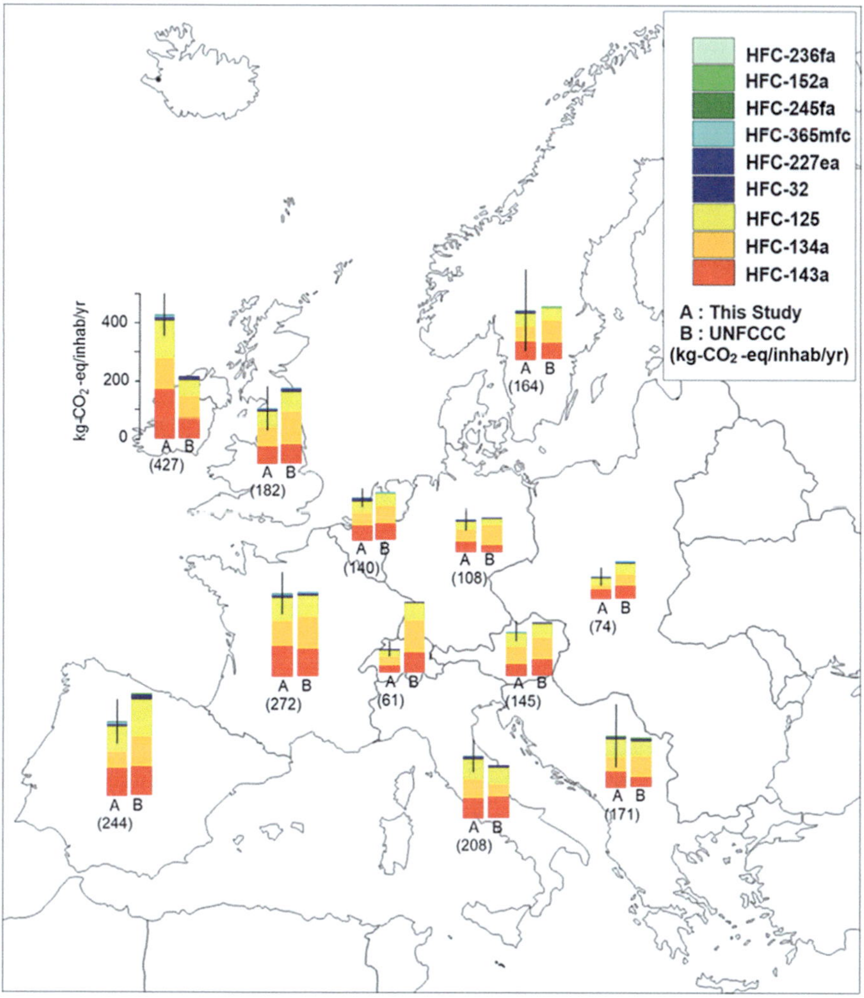

Fig. 4 Per capita HFC emissions from 12 macro-areas in the European geographic domain. Emissions, given in kg-CO_2-eq·year^{-1} ·inhabitants^{-1}, are averaged over 2008–2014. Column A, Bayesian inversion results; Column B, UNFCCC country reports (Adapted from Graziosi et al. 2017)

Other fluorinated gases monitored at CMN include PFC-218, PFC-318, and SO_2F_2, whose time series and trends are reported in Fig. 3 and Tables 4a, 4b, respectively.

Methane and Nitrous Oxide CH_4 measurements at CMN started in 2008, with an average concentration over 2008–2016 of 1881.5 ppb and an increasing trend of 9.3 ppb year^{-1}. CH_4 data have been incorporated in two different studies focused on the European scale. Weaver et al. (2014) investigated the sensitivity of future spaceborne lidar measurements to changes in surface CH_4 emissions, using surface CH_4 observations from nine European sites and a Lagrangian transport model to infer emissions for 2010. They found that, consistently with the main CH_4 sources, strongest emissions were from the Netherlands, the coal mines in Poland, and wetlands in Southern Finland. The final aim of the study was to test whether proposed CH_4 lidar instruments were sensitive to changes in surface emissions. More recently, Tsuruta et al. (2016) estimated global methane emissions for 2000–2012 using the CarbonTracker Europe-CH_4 (CTE-CH_4) data assimilation system. The estimates were evaluated against assimilated in situ atmospheric CH_4 observations. They estimated global total a posteriori emissions for 2000–2012 of 515–517 Tg year^{-1} increasing by about 18–19 Tg year^{-1} from 2001–2006 to 2007–2012. For what the European emissions are concerned, no significant trend in anthropogenic or biospheric emissions was found in Europe as a whole. However, the estimates were higher than those given in the EDGARv4.2 FT2010 inventory (EDGAR 2011) for southern Europe and lower in northern Europe.

N_2O is measured at CMN since 2008. Average concentrations during 2008–2016 are 326.5 ppb, with an increasing trend of 0.8 ppb year^{-1}, with that the same growth rate given for the global scale by Carpenter et al. (2014).

4 Conclusions

High-quality atmospheric measurements of radiatively active gases have been crucial in revealing the growth of atmospheric levels of these environmentally relevant species.

The research activity on non-CO_2 GHGs at CMN includes the observation of a wide range of compounds that are measured continuously at the site, in most cases, since more than a decade. The inclusion of CMN in important global networks makes these data comparable with those measured in other sites across the globe and allows the inclusion of these high-quality data in studies that are interested both at the global and at the regional (European) scale. The analysis of these data using inverse modeling approaches has allowed us to provide emission estimates at the regional scale that constitute an important support to improve bottom-up emission data that each country is required to submit every year in the frame of the most important international global protocols aimed at combatting climate change. Such activity is within the scopes of the Integrated Global Greenhouse Gas Information System (IG^{3}IS) of the World Meteorological Organization, a framework capable of accepting and promoting a range of advanced GHG emission quantification capabilities in order to improve the quality of and confidence in GHG emission inventories.

Acknowledgments The results described in this chapter have been obtained, thanks to the collaboration with the science teams of the AGAGE consortium. This research started under the EU FP5 Project SOGE (EVK2-2000-00674). The InGOS EU FP7 Infrastructure project (grant agreement n° 284274) supported the observation and calibration activities. The University Consortium Consorzio Interuniversitario Nazionale per la Fisica delle Atmosfere e delle Idrosfere (CINFAI) supported F. Graziosi grant (RITMARE Flagship Project). The "O. Vittori" observatory is supported by the National Research Council of Italy and the Italian Ministry of Education, University, and Research, through the Project of National Interest "Nextdata."

References

Bergamaschi P, Behrend H, Jol A (eds) (2004) Inverse modelling of national and EU greenhouse gas emission inventories – report of the workshop inverse modelling for potential verification of national and EU bottom-up GHG inventories, EN/ISBN 92-894-7455-6, European Commission Joint Research Centre, Ispra

Bousquet P, Ciais P, Miller JB et al (2006) Contribution of anthropogenic and natural sources to atmospheric methane variability. Nature 443:439–443

Bruhwiler L, Dlugokencky E, Masarie K et al (2014) CarbonTracker-CH4: an assimilation system for estimating emissions of atmospheric methane. Atmos Chem Phys 14:8269–8293

Brunner D, Henne S, Keller CA et al (2012) An extended Kalman-filter for regional scale inverse emission estimation. Atmos Chem Phys 12:3455–3478

Carpenter LJ, Reimann S (Lead Authors) et al (2014) Ozone-Depleting Substances (ODSs) and other gases of interest to the Montreal Protocol, Chapter 1 in Scientific Assessment of Ozone Depletion, Global Ozone Research and Monitoring Project - Report no. 55, World Meteorological Organization, Geneva

Dlugokencky EJ, Masarie KA, Lang PM et al (1998) Continuing decline in the growth rate of the atmospheric methane burden. Nature 393:447–450

Dlugokencky EJ, Bruhwiler L, White JWC et al (2009) Observational constraints on recent increases in the atmospheric CH_4 burden. Geophys Res Lett 36:L18803

Eckhardt S, Prata AJ, Seibert P et al (2008) Estimation of the vertical profile of sulfur dioxide injection into the atmosphere by a volcanic eruption using satellite column measurements and inverse transport modelling. Atmos Chem Phys 8:3881–3897

EDGAR (Emission Database for Global Atmospheric Research), release version 4.2 FT2010 (2011) European Commission, Joint Research Centre (JRC)/Netherlands Environmental Assessment Agency (PBL): http://edgar.jrc.ec.europa.eu

Fang X, Velders GJ, Ravishankara M et al (2016) Hydrofluorocarbon (HFC) emissions in China: an inventory for 2005–2013 and projections to 2050. Environ Sci Technol 50:2027–2034

Farman JC, Gardiner G, Shanklin D (1985) Large losses of total ozone in Antarctica reveal seasonal ClO_x/NO_x interaction. Nature 315:207–210

Fraser A, Palmer PI, Feng L et al (2013) Estimating regional methane surface fluxes: the relative importance of surface and GOSAT mole fraction measurements. Atmos Chem Phys 13:5697–5713

Fraser PJ, Dunse B, Manning AJ (2014) Australian carbon tetrachloride (CCl_4) emissions in a global context. Environ Chem 11:77–88

Ghosh A, Patra PK, Ishijima K et al (2015) Variations in global methane sources and sinks during 1910–2010. Atmos Chem Phys 15:2595–2612

Giostra U, Furlani F, Arduini J et al (2011) The determination of a regional atmospheric background mixing ratio for anthropogenic greenhouse gases: a comparison of two independent methods. Atmos Environ 45:7396–7405

Graziosi F, Arduini J, Furlani F et al (2015) European emissions of HCFC-22 based on eleven years of high frequency atmospheric measurements and a Bayesian inversion method. Atmos Environ 112:196–207

Graziosi F, Arduini J, Bonasoni P et al (2016) Emissions of carbon tetrachloride from Europe. Atmos Chem Phys 16:12849–12859

Graziosi F, Arduini J, Furlani F et al (2017) European emissions of the powerful greenhouse gases hydrofluorocarbons inferred from atmospheric measurements and their comparison with annual national reports to UNFCCC. Atmos Environ, accepted

Greally BR, Manning AJ, Reimann S et al (2007) Observations of 1,1-difluoroethane (HFC-152a) at AGAGE and SOGE monitoring stations in 1994–2004 and derived global and regional emission estimates. J Geophys Res 112:D06308

Henne S, Brunner D, Oney B et al (2016) Validation of the Swiss methane emission inventory by atmospheric observations and inverse modelling. Atmos Chem Phys 16:3683–3710

Houweling S, Krol M, Bergamaschi P et al (2014) A multi-year methane inversion using SCIAMACHY, accounting for systematic errors using TCCON measurements. Atmos Chem Phys 14:3991–4012

Hu L, Montzka SA, Miller JB et al (2015) U.S. emissions of HFC-134a derived for 2008–2012 from an extensive flask-air sampling network. J Geophys Res Atmos 120:801–825

Hu L, Montzka SA, Miller BR et al (2016) Continued emissions of carbon tetrachloride from the U.S. nearly two decades after its phase-out for dispersive uses. Proc Natl Acad Sci U S A 113:2880–2885

IPCC (Intergovernmental Panel on climate change) (2001) climate change 2001: the scientific basis. Contribution of working group I to the third assessment report. Houghton JT, Ding Y, Griggs DJ, Noguer M, van der Linden PJ, Dai X, Maskell K, Johnson CA (eds) Cambridge University Press, Cambridge/New York

IPCC (Intergovernmental Panel on climate change) (1990) Report prepared by working group I. Houghton JT, Jenkins GJ, Ephraums JJ (eds) Cambridge University Press, Cambridge/New York/Melbourne

IPCC (Intergovernmental Panel on Climate Change) (2005) IPCC/TEAP (technology and economic assessment Panel) special report on safeguarding the ozone layer and the global climate e system: issues related to Hydrofluorocarbons and Perfluorocarbons. Prepared by working group I and III of the Intergovernmental Panel on Climate Change, and the Technology and Economic Assessment Panel, Cambridge University Press, Cambridge

IPCC (Intergovernmental Panel on Climate Change) (2006) Guidelines for National Greenhouse Gas Inventories, Prepared by the National Greenhouse Gas Inventories Programme, Eggleston HS, Buendia L, Miwa K et al (eds) Published: IGES, Japan

IPCC (Intergovernmental Panel on Climate Change) (2013) Summary for Policymakers. In: Climate Change 2013: The Physical Science Basis. Contribution of working group I to the Fifth Assessment Report of the Intergovernmental Panel on Climate Change [Stocker TF, Qin D, Plattner G-K et al (eds) Cambridge University Press, Cambridge/New York

Keller CA, Brunner D, Henne S et al (2011) Evidence for under-reported western European emissions of the potent greenhouse gas HFC-23. Geophys Res Lett 38:L15808

Keller CA, Hill M, Vollmer MK et al (2012) European emissions of halogenated greenhouse gases inferred from atmospheric measurements. Environ Sci Technol 46:217–225

Kim J, Li S, Kim K-Y et al (2010) Regional atmospheric emissions determined from measurements at Jeju Island, Korea: halogenated compounds from China. Geophys Res Lett 37:L12801

Kirschke S, Bousquet P, Ciais P et al (2013) Three decades of global methane sources and sinks. Nat Geosci 6:813–823

Krol MC, Lelieveld J, Oram DE et al (2003) Continuing emissions of methyl chloroform from Europe. Nature 421:131–135

Li S, Kim J, Kim K-R et al (2010) Emissions of halogenated compounds in East Asia determined from measurements at Jeju Island, Korea. Environ Sci Technol 45:5668–5675

Lunt MF, Rigby M, Ganesan AL (2015) Reconciling reported and unreported HFC emissions with atmospheric observations. Proc Natl Acad Sci U S A 112:5927–5593

Maione M, Arduini J, Mangani G et al (2004) Evaluation of an automated gas chromatographic - mass spectrometric instrumentation to be used for continuous monitoring of trace anthropogenic greenhouse gases. Int J Environ Anal Chem 84(4):241–253

Maione M, Giostra U, Arduini J et al (2013) Ten years of continuous observations of stratospheric ozone depleting gases at Monte Cimone (Italy) - comments on the effectiveness of the Montreal Protocol from a regional perspective. Sci Tot Environ 445-446:155–164

Maione M, Graziosi F, Arduini J et al (2014) Estimates of European emissions of methyl chloroform using a Bayesian inversion method. Atmos Chem Phys 14:9755–9770

Manning AJ, O'Doherty S, Jones AR et al (2011) Estimating UK methane and nitrous oxide emissions from 1990 to 2007 using an inversion modelling approach. J Geophys Res 116(D2):D02305

Meirink JF, Bergamaschi P, Krol MC (2008) Four-dimensional variational data assimilation for inverse modelling of atmospheric methane emissions: method and comparison with synthesis inversion. Atmos Chem Phys 8:6341–6353

Miller BR, Weiss RF, Salameh PK et al (2008) Medusa: a sample preconcentration and GC/MS detector system for in situ measurements of atmospheric trace halocarbons, hydrocarbons, and sulphur compounds. Anal Chem 80:1536–1545

Miller BR, Rigby M, Kuijpers LJM et al (2010) HFC-23 (CHF_3) emission trend response to HCFC-22 ($CHClF_2$) production and recent HFC-23 emission abatement measures. Atmos Chem Phys 10:7875–7890

Miller JB, Lehman SJ, Montzka SA et al (2012) Linking emissions of fossil fuel CO_2 and other anthropogenic trace gases using atmospheric 14 CO_2. J Geophys Res 117:D08302

Molina MJ, Rowland FS (1974) Stratospheric sink for chlorofluoromethanes: chlorine atom-catalysed destruction of ozone. Nature 249:810–812

Montzka SA, Spivakovsky CM, Butler JH et al (2000) New observational constraints for atmospheric hydroxyl on global and hemispheric scales. Science 288:500–503

Montzka SA, Kuijpers L, Battle MO et al (2010) Recent increases in global HFC-23 emissions. Geophys Res Lett 37:L02808

Montzka SA, Reimann S (lead authors) et al (2011) Ozone-Depleting Substances (ODSs) and related chemicals, Chapter 1 in Scientific Assessment of Ozone Depletion: 2010, Global Ozone Research and Monitoring Project-Report No 52, World Meteorological Organization, Geneva

Nisbet E, Weiss R (2010) Top-down versus bottom-up. Science 328:1241–1243

O'Doherty SJ, Nickless G, Bassford M et al (1999) Separation of hydrohalocarbons and chlorofluorocarbons using a cyclodextrin gas solid chromatography capillary column. J Chromatogr A 832:253–258

O'Doherty S, Rigby M, Mühle J et al (2014) Global emissions of HFC-143a (CH_3CF_3) and HFC-32 (CH_2F_2) from in situ and air archive atmospheric observations. Atmos Chem Phys 14:9249–9258

Odabasi M, Elbir T, Dumanoglu Y et al (2014) Halogenated volatile organic compounds in chlorine-bleach-containing household products and implications for their use. Atmos Environ 92:376–383

Prinn RG, Weiss RF, Fraser PJ et al (2000) A history of chemically and radiatively important gases in air deduced from ALE/GAGE/AGAGE. J Geophys Res 105(D14):17751–17792

Prinn RG, Huang J, Weiss RF et al (2001) Evidence for substantial variations of atmospheric hydroxyl radicals in the past two decades. Science 292:1882–1888

Prinn RG, Huang J, Weiss RF et al (2005) Evidence for variability of atmospheric hydroxyl radicals over the past quarter century. Geophys Res Lett 32:L07809

Reimann S, Manning AJ, Simmonds PG et al (2005) Low European methyl chloroform emissions inferred from long term atmospheric measurements. Nature 433:506–508

Rigby M, Prinn RG, Fraser PJ et al (2008) Renewed growth of atmospheric methane. Geophys Res Lett 35:L22805

Saikawa E, Rigby M, Prinn RG et al (2012) Global and regional emissions estimates for HCFC-22. Atmos Chem Phys 12:10033–10050

Seibert P (2000) Inverse modelling of sulphur emissions in Europe based on trajectories. In: Kasibhatla P, Heimann M, Rayner P et al (eds) Inverse Methods in Global Biogeochemical Cycles, Geophysical Monograph, vol 114. American Geophysical Union, Washington, DC, pp 147–154. ISBN:0-87590-097-6

Seibert P (2001) Inverse modelling with a Lagrangian particle dispersion model: application to point releases over limited time intervals. In: Schiermeier FA, Gryning SE (eds) Air pollution modeling and its application XIV. Kluwer Academic Publishers, Dordrecht

Shindell D, Kuylenstierna JCI, Vignati E et al (2012) Simultaneously mitigating near-term climate change and improving human health and food security. Science 335:183–189

Simmonds PG, Cunnold DM, Weiss RF et al (1998) Global trends and emissions estimates of CCl4 from in-situ background observations from July 1978 to June 1996. J Geophys Res 103:16,017–16,027

Simmonds PG, O'Doherty S, Derwent RG et al (2004) AGAGE observations of methyl bromide and methyl chloride at the Mace Head, Ireland and Cape Grim, Tasmania, 1998–2001. J Atmos Chem 47(3):243–269

Simmonds PG, Rigby M, Manning AJ et al (2016) Global and regional emissions estimates of 1,1-difluoroethane (HFC-152a, CH_3CHF_2) from in situ and air archive observations. Atmos Chem Phys 16:365–382

SPARC (Stratosphere-troposphere Processes And their Role in Climate) (2016) SPARC report on the mystery of carbon tetrachloride. Liang Q, Newman PA, Reimann S (eds) SPARC report no. 7, WCRP-13/2016 doi: 10.3929/ethz-a-010690647

Stemmler K, Folini D, Ubl S et al (2007) European emissions of HFC-365mfc, a chlorine-free substitute for the foam blowing agents HCFC-141b and CFC-11. Environ Sci Technol 41:1145–1151

Stohl A, Forster C, Frank A et al (2005) Technical note: the Lagrangian particle dispersion model FLEXPART version 6.2. Atmos Chem Phys 5:2461–2474

Stohl A, Seibert P, Arduini J et al (2009) An analytical inversion method for determining regional and global emissions of greenhouse gases: sensitivity studies and application to halocarbons. Atmos Chem Phys 9:1597–1620

Stohl A, Kim J, Li S et al (2010) Hydrochlorofluorocarbon and hydrofluorocarbon emissions in East Asia determined by inverse modelling. Atmos Chem Phys 10:3545–3560

Tsuruta A, Aalto T, Backman L et al (2016) Development of CarbonTracker Europe-CH4 – Part 2: global methane emission estimates and their evaluation for 2000–2012. Geoscientific Model Development Discussions doi:10.5194/gmd-2016-182, in review, 2016

UNEP (United Nations Environ Programme) (1987) The Montreal protocol on substances that deplete the ozone layer. United Nations Environ Programme, Nairobi

UNEP (United Nations Environ Programme) (2007) Report of the halons technical options committee, 2006 assessment. United Nations Environment Programme, Ozone Secretariat, Nariobi, Kenya

UNEP (United Nations Environ Programme) (2011) HFCs: a critical link in protecting climate and the ozone layer. United Nations Environment Programme, Nairobi

UNEP (United Nations Environ Programme) (2014) Report of the methyl bromide technical options committee. United Nations Environment Programme, Nairobi

UNEP (United Nations Environ Programme) (2015a) Proposed amendment to the Montreal Protocol submitted by Canada, Mexico and the United States of America. United Nations Environment Programme, Nairobi

UNEP (United Nations Environ Programme) (2015b) Proposed amendment to the Montreal Protocol submitted by European Union and its member states. United Nations Environment Programme, Nairobi

UNEP (United Nations Environ Programme) (2015c) Proposed amendment to the Montreal Protocol submitted by India. United Nations Environment Programme, Nairobi

UNEP (United Nations Environ Programme) (2015d) Proposed amendment to the Montreal Protocol submitted by Kiribati, Marshall Islands, Mauritius, Micronesia (federated states of), Palau, Philippines, Samoa and Solomon Islands. United Nations Environment Programme, Nairobi

UNFCCC (United Nations Framework Convention on Climate Change) (1997) Kyoto Protocol to the United Nations framework Convention on climate change. United Nations, Geneva

Velders GJM, Andersen SO, Daniel JS et al (2007) The importance of the Montreal Protocol in protecting climate. Proc Natl Acad Sci U S A 104(12):4814–4819

Velders GJM, Fahey DW, Daniel JS et al (2009) The large contribution of projected HFC emissions to future climate forcing. Proc Natl Acad Sci U S A 106:10949–10954

Velders GJM, Ravishankara AR, Miller MK et al (2012) Preserving Montreal protocol climate benefits by limiting HFCs. Science 335:922–923

Velders GJM, Fahey DW, Daniel JS et al (2015) Future atmospheric abundances and climate forcings from scenarios of global and regional hydrofluorocarbon (HFC) emissions. Atmos Environ 123:200–209

Vollmer MK, Zhou LX, Greally BR et al (2009) Emissions of ozone-depleting halocarbons from China. Geophys Res Lett 36:L15823

Vollmer MK, Miller BR, Rigby M et al (2011) Atmospheric histories and global emissions of the anthropogenic hydrofluorocarbons HFC-365mfc, HFC-245fa, HFC-227ea, and HFC-236fa. J Geophys Res 116:D08304

Vollmer MK, Mühle J, Trudinger CM et al (2016) Atmospheric histories and global emissions of halons H-1211 ($CBrClF_2$), H-1301 ($CBrF_3$), and H-2402 ($CBrF_2CBrF_2$). J Geophys Res Atmos 121(7):3663–3686

Weaver C, Kiemle C, Kawa SR et al (2014) Retrieval of methane source strengths in Europe using a simple modeling approach to assess the potential of spaceborne lidar observations. Atmos Chem Phys 14:2625–2637

Weiss RF, Prinn RG (2011) Quantifying greenhouse-gas emissions from atmospheric measurements: a critical reality check for climate legislation. Phil Trans R Soc A 369:1925–1942

Xiao X, Prinn RG, Fraser PJ et al (2010a) Atmospheric three-dimensional inverse modelling of regional industrial emissions and global oceanic uptake of carbon tetrachloride. Atmos Chem Phys 10:10421–10434

Xiao X, Prinn RG, Fraser PJ et al (2010b) Optimal estimation of the surface fluxes of methyl chloride using a 3-D global chemical transport model. Atmos Chem Phys 10:5515–5533

Xu Y, Zaelke D, Velders GJM et al (2013) The role of HFCs in mitigating 21st century climate change. Atmos Chem Phys 13:6083–6089

Yao B, Vollmer MK, Zhou LX et al (2012) In-situ measurements of atmospheric hydrofluorocarbons (HFCs) and perfluorocarbons (PFCs) at the Shangdianzi regional background station, China. Atmos Chem Phys 12:10181–10193

Yokouchi Y, Taguchi S, Saito T et al (2006) High frequency measurements of HFCs at a remote site in east Asia and their implications for Chinese emissions. Geophys Res Lett 33:L21814

Investigation of Atmospheric Reactive Gases at Mt. Cimone

Abstract The Mediterranean basin represents a global hot spot for climate change, air quality, and anthropogenic contributions to these issues. In this region, a relevant role is played by reactive gases (RGs), i.e., O_3, CO, SO_2, NO_x, and VOCs. At the "O. Vittori" observatory at Mt. Cimone, a number of RGs are observed since more than 20 years. These observations are carried out within the framework of international program (e.g., WMO/GAW, ACTRIS) representing a valuable resource of information for quantifying the variability of these atmospheric key compounds and for investigating the influence of specific atmospheric processes to this variability (e.g., polluted air mass transport, open fire emissions, mineral dust outbreaks, stratospheric intrusion events). These observations were used to experimental applications like near-real-time data delivery services or model verification and satellite data ground-truthing.

Keywords Mt. Cimone • Reactive gases • STE • Biomass burning • Anthropogenic pollution • Saharan dust • Mediterranean basin

1 Introduction

Reactive gases (RGs) are a group of atmospheric molecules with short lifetimes (ranging from hours to a few months). Among them, important roles are played by tropospheric ozone (O_3), carbon monoxide (CO), reactive nitrogen gases (NO_y), volatile organic compounds (VOC), and reactive sulfur gases (SO_x). In particular, the RGs trigger the oxidative capacity of the atmosphere as they represent the key source of reactive atomic and free radical species (e.g., O*, OH, HO_2, RO_2). By this way, they contribute in defining the cleansing capacity of the atmosphere, by initiating the oxidation processes which affect the variability of greenhouse gases (see previous Chapter) and air pollutants. By means of their impact on greenhouse gases, RGs have an indirect role on atmosphere radiative forcing and thus on the Earth's climate. Moreover, O_3 is also a powerful greenhouse gas, thus playing a direct role on climate (especially at regional scale). Due to its shorter atmospheric lifetime in respect to other "well-mixed" greenhouse gases (like CO_2, N_2O, or CFCs), it is also indicated as "short-lived climate forcer/pollutant" (SLCF/P, see UNEP and WMO

P. Cristofanelli et al., *High-Mountain Atmospheric Research*, SpringerBriefs in Meteorology, DOI 10.1007/978-3-319-61127-3_3

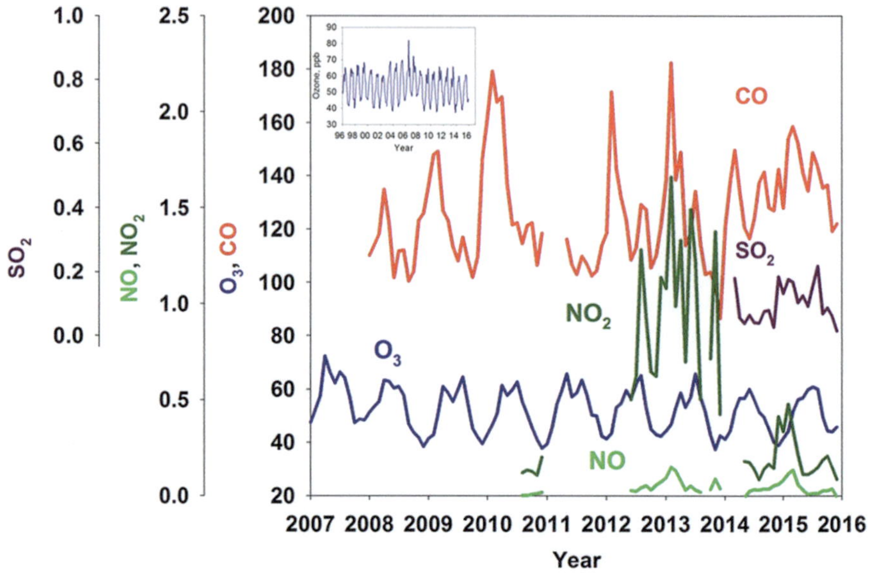

Fig. 1 Time plot of monthly mean values of O_3, CO, NO, NO_2, and SO_2 at CMN. The insert reports the whole time series for O_3

2011). Now, it is widely recognized that the mitigation of O_3 precursor emissions (especially CH_4) is a very effective way to limit the global warming for this century under the "safety" threshold of +2 °C from the preindustrial age (UNEP and WMO 2011). Moreover, NOx, VOC, and SO_2 also influence climate by affecting the formation of aerosol particles and clouds (Schultz et al. 2015).

Also, RGs are important contributors to air pollution. They have direct adverse effect on human health and ecosystems, and they participate to the formation of photochemical smog or secondary aerosol formation. In this regard, the Mediterranean basin, where CMN is located, represents a global hot spot for the formation of photochemical ozone (O_3) due to the availability of natural and anthropogenic precursors (Monks et al. 2009).

As reported by Schultz et al. (2015), information of background RG spatial distribution and variability is essential for assessing the effectiveness of emission mitigation measures. To this aim, a potentially effective role is provided by continuous measurements of RGs at high mountain region, far from direct source of pollution and representative of the background tropospheric conditions. In respect to other locations, mountain sites have the advantage to provide measurements with higher spatial representativeness (Henne et al. 2010), especially under specific conditions (e.g., without the influence of vertical transport regimes bringing air masses from their local PBL). This makes mountain sites, very suitable and strategic locations where performing continuous RG measurements, both to disentangle underlying processes affecting their background variability and to contribute to the interpretation of global distribution and trends (Fig. 1).

Table 1 List of RGs observed at CMN with starting year of measurements, typical range of observed levels, and their principal role in the climatic system and atmospheric chemistry

Reactive gas	Year	Typical level at CMN (ppb)	Role in the climate system and atmospheric chemistry
Ozone (O_3)	1996	$10 \div 10^2$ ppb	GHG, component of photochemical smog, oxidant
Carbon monoxide (CO)	2007	~10^2 ppb	Ozone precursor
Nitrogen oxides (NO, NO_2)	2013	$10^{-2} \div 1$ ppb	Component of photochemical smog, precursor of ozone and secondary aerosol
Non-methane volatile Organic compounds (NMVOC)	2010	$10^0 \div 10^2$ ppt	Component of photochemical smog, precursor of ozone, CO, and secondary aerosol
Sulfur dioxide (SO_2)	2015	$10^{-1} \div 1$ ppb	Precursor of secondary aerosol, air pollutant
Columnar NO_2	1993		Component of photochemical smog, precursor of ozone and secondary aerosol

Measurements of RGs are challenging, especially at remote measurement sites characterized by harsh environmental conditions like mountains. Indeed, at these locations, their mole fractions are typically in the range of nmol/mol (i.e., ppb) to pmol/mol (i.e., ppt). Therefore, sophisticated instrumentation, often not directly available through the commercial market, is needed to quantify their abundance with specific (low) uncertainties. Furthermore, RGs are often difficult to sample just because of their reactivity. Thus, sampling systems must be able not only to guarantee the correct execution of air sampling during ice rimming or windy conditions (typical for high mountain sites) but also to minimize interferences with air sample and the variation of chemical properties during the sampling phase. Further critical points for execution of RG observations are the maintaining of consistent calibration system: some of the reference materials are unstable and need frequent replacement. Specific materials must be used to avoid artifacts. As stressed by Schultz et al. (2015), the long-term trends of reactive gases are generally small relative to their atmospheric variability. This need very high levels of data quality objectives besides the ability (not only technical and scientific but also financial) to sustain continuous long-term observations.

At the CMN "O. Vittori" observatory, a suite of RGs is continuously monitored since 1996 when surface O_3 investigation was started in the framework of the VOTALP (Vertical Ozone Transport in the Alps) EU Project. From this year, the number of measurement programs increased almost continuously (Table 1): today, more than 17 RGs are continuously measured at this mountain site. In the following, we provided a description of the experimental setup, technical details, and typical average mixing ratios and variability of RGs observed at CMN. Then we provide an overview of the principal scientific studies and applications of RG observations at CMN. Open data on RGs collected at CMN from 1996 to 2015 can be found at the

World Data Center for Greenhouse Gases (WDCGG) of WMO/GAW (http://ds.data.jma.go.jp/gmd/wdcgg/) and at the World Data Center for Reactive Gases (WDCRG) of WMO/GAW (http://ebas.nilu.no).

2 Method

2.1 Surface Ozone (O_3)

At CMN, the surface O_3 measurements have been carried out from 1996 by using a UV-absorption analyzer Dasibi 1108 W/GEN. Starting from 2012 a Thermo Tei49i instrument was used at the station. After one full year of intercomparison, the latter instrument has been designed as the "master" O_3 analyzer at CMN. QA/QC strategy includes daily automatic execution of zero and span checks, as well as multipoint calibrations every 3 months. The zero air was generated by using an activated charcoal cartridge, while an internal UV source was used to generate a span level of approximately 100 ppb. Since 1996, the UV analyzer was regularly calibrated (roughly every 3 months) with the laboratory transfer standard (Dasibi 1008PC until 2012, hereinafter Thermo 49iPS). On October 2006, the Dasibi 1008PC was compared against the IMGC-O3SRP primary standard at the National Institute of Metrological Research (INRIM) in Turin, Italy (INRIM calibration certificate N. 06-1009-01): an expanded uncertainty of 2 nmol/mol was assessed in the measurement range 0–100 nmol/mol. Starting from June 2008, every 2 years, the laboratory standard was intercompared against the SRP#15 primary standard hosted at WCC-Empa: an expanded uncertainty lower than 2.6 nmol/mol was assessed in the measurement range 0–100 nmol/mol. Before 2006 the laboratory transfer standard was routinely calibrated by the manufacturer.

Only from November 2002 to January 2003, a major data gap exists in the CMN surface O_3 data series due to problem with the blower of the sampling system.

2.2 Carbon Monoxide (CO)

Starting from February 2008, CO mixing ratio was measured with a custom gas chromatograph equipped with a reduction gas detector (GC-RGD), Trace Analytical RGD2. Every 15 min, an air sample is injected into the gas chromatograph for separation and then analyzed for CO via mercury oxidoreduction and detection of mercury vapors by UV absorption. Each analysis was alternated with a calibration sample by means of working standards in real air matrix representative for ambient mixing ratios. This guarantees a continuous check of the detector calibration (Novelli 1999) with an accuracy, in terms of daily relative standard deviation over repeated analysis of the working standards (48 runs) of 0.72% and a total expanded uncertainty of 1.51% ($k = 2$).

Since 2008, a customized GC-FID (Agilent 6890 N) is run at CMN. CO determination is performed after reduction to CH_4 through a hot nickel catalyzer. Separation is performed on a double column system (precolumn/backflush: Unibeads 1 s; analytical column: Molecular Sieve 5A). Sample and calibration strategies are the same than for the GC-NDIR system but with injection of ambient air sample every 30 min. Length of run is 15 min. Each ambient air sample is bracketed with calibration injections. From 2008 to 2012, the working standards were prepared at Max Planck Institute for Biogeochemistry in Jena and indirectly referenced against NOAA 2004 scale. Since 2012, a set of standards provided by NOAA is available at CMN. The sensitivity is generally below 2 ppb, while the measurement detection limit, defined as the theoretical LOQ ($S/N = 10$), was estimated to 45 ppb. The two instruments ran in parallel from 2008 to 2012. RGD detectors are characterized by nonlinear response. To test this nonlinearity, the GC-RGD system was intercompared with the GC-FID. Even if only for the limited range of the typical mixing ratios observed at CMN, a correction function was obtained.

Besides GC systems, CO is measured by nondispersive infrared (NDIR) technique at CMN since 2012. The system is based on a Thermo Tei48C-TL analyzer which uses gas filter correlation technology for determining CO ambient concentration. Ambient air is drawn in the instrument using the internal pump at 1.0 l min^{-1}, and the instrument is connected to the laboratory manifold by approx. 1.5 m of Teflon tube. With the aim of minimizing possible influence of water vapor in the NDIR detection, the ambient air passes through a drying system (Nafion© dryer) and is then injected in the measurement cell. The system is also equipped with a purge air unit (constituted of a Parker drying system and steel tube filled with Sofnocat 423) for flushing the gas correlation wheel to reduce the possible interference due to contaminants and therefore increasing measurement reliability.

The Thermo 48C-TL instrument is characterized by a strong drift of the zero value due to changes in ambient (room) temperature: to minimize the influence of temperature on the measurements, a specific software was designed and used to control the zero calibration, forcing the instrument to execute this calibration every 30 min by using dry air scrubbed by Sofnocat 423. The zero air is then injected in the cell for 15 min, and only the final 5 min are averaged to produce the zero value that is applied to the instrument using our self-built software.

A span calibration is performed once a day, starting at 00:30 AM and ending at 00:45 AM. The span value is 500 nmol/mol, and this concentration is obtained from the dilution of a 10 μmol/mol CO standard cylinder (Producer: Messer Italia) with zero air produced by a zero air generator. Every 3 months, a multipoint calibration is carried out by using a set of six NOAA standards. In this way, the measurements are referred to the WMO CO_X2014A calibration scale.

As assessed during the audit performed at CMN on September 2012 by the GAW World Calibration Center (Zellweger et al. 2012), the expanded uncertainties associated to a CO mixing ratio X^{CO} (GC/FID and NDIR, respectively) were assessed to be:

$$u_{CO}^{CG/FID}(\text{ppb}) = \text{sqrt}\left(2.6\,\text{ppb}^2 + 1.01\text{e}-04^* X_{CO}^{\ 2}\right)$$

$$u_{CO}^{NDIR}(\text{ppb}) = \text{sqrt}\left(136.9\,\text{ppb}^2 + 1.01\text{e}-04^* X_{CO}^{\ 2}\right)$$

2.3 Nitrogen Oxides (NO and NO_2)

NO_x observations are continuously carried out at CMN since June 2014 by using a chemiluminescence analyzer (Thermo Tei42i-TL) equipped with a photolytic converter (Blue Light Converter by Air Quality Design). Every 48 h, zero and span checks are carried out for NO by using an external zero air source (Thermo 1160 dry air generator equipped with Purafill© and active charcoal scrubbers) and dilution of certified NO mixture in N_2 (5 ppm ± 3.5%). To determine the efficiency of the NO_2 photolytic converter, gas phase titration (GPT) is used to titrate about 80% of the NO obtained by dilution. The obtained calibration coefficients (zero, span, and conversion efficiency) were used to correct the obtained raw mixing ratios. To evaluate instrument linearity over a range from 1 to 20 ppb, for NO, multipoint calibrations by dynamic dilution are carried out on a yearly basis. In March 2013, the NO_x instrument participated to a round robin test executed in the framework of the ACTRIS EU Project. A detection limit (zero noise) of 0.06 ppb and 0.14 ppb was determined for NO and NO_2, respectively.

During the period 2011–2014, NO_x was measured by using a Philips TE42 analyzer which was equipped with a heated (300 °C) molybdenum converter for NO_2 determination. As indicated by Steinbacher et al. (2007), NO_2 reading obtained in this way can be overestimated up to ~50% due to the interference of other oxidized nitrogen compounds such as peroxyacetyl nitrate (PAN) and nitric acid. Thus great caution must be deserved in comparing the NO_x data series obtained by these different experimental setups.

2.4 Sulfur Dioxide (SO_2)

The SO_2 instrument (Thermo 43iTLE) is based on the UV fluorescence technology. Daily execution of zero/span checks were carried out by an internal span source (permeation tube with emission rate of 44 ng/min ± 10%) and external zero source (activated charcoal-filled cartridge). A dilution system (Environics 4000) is used to perform every 6 months multipoint calibration (range, 0.5–10 ppb) from a certified standard of 235 ppb SO_2 in N_2 (s/n: 12048642) provided by SIAD (expanded uncertainty with coverage factor k = 235 ppb). A detection limit of 0.11 ppb was experimentally assessed on a 1-minute basis by evaluating the zero noise.

Table 2 RSD, LODs, and LOQs for the 12 NMVOCs measured at CMN since 2010

Species	%RSD	TU ($k = 2$) (ppt)	LOD (ppt)	LOQ (ppt)
Ethyne	1.9	12.3	1.0	3.3
Propane	2.9	23.2	1.0	3.3
Benzene	0.7	6.8	1.0	3.3
i-butane	1.5	4.1	1.0	3.3
n-butane	1.3	12.6	1.0	3.3
i-pentane	1.3	3.8	1.0	3.3
n-pentane	2.6	4.8	1.0	3.3
Toluene	1.3	3.3	1.0	3.3
Ethylbenzene	1.7	2.8	1.0	3.3
o-xylene	1.9	2.8	1.0	3.3
m + p xylene	2.2	4.8	1.0	3.3

2.5 *Non-methane Volatile Organic Compounds (NMVOCs)*

Since 2010, observations of 12 different NMVOCs in ambient air are carried out by gas chromatography-mass spectrometry (GC-MS) preceded by online sample enrichment using adsorbent material. To this aim, a GC-MS Agilent 6850–5975 has been equipped with auto-sampling/pre-concentration device (Markes International, UNITY2-Air Server2). The pre-concentration is achieved at low temperature (−30 °C): for this reason the sample (for both real air and working standards) is dried by a Nafion© tube (for more details, see Lo Vullo et al. 2016a). Table 2 reports instrumental performances. Annual mean relative standard deviations (RSD) are evaluated from the repeated working standard measurements. Absolute total uncertainty (TU, coverage factor = 2) is calculated including the certified uncertainty in the standard gas mixing ratio. Limits of detection (LODs), signal-to-noise ratio ($S/N > 3$), and limits of quantitation, LOQs ($S/N > 10$), have been indirectly estimated on the chromatogram of the working standard run.

Every 2 h an ambient air sample is injected and analyzed (total time of the analysis cycle, 26 min). The air sampling is bracketed by working standard analyses, in order to detect and correct short-term instrumental drift. These working standards are composed by real air which is pumped during relatively clean-background conditions into 35-l electropolished stainless steel canisters (Essex Cryogenics, Missouri, USA) using a modified oil-free compressor (SA-3 or SA-6Rix California, USA) up to ~60 bar. The canisters are humidified during the pumping process to improve the stability of the compounds (Miller et al. 2008) and to ensure similarity with the real air matrix. These working standards are referred against a reference standard by NPL (National Physical Laboratory, Teddington, UK). The NPL gas mixture contains 30 hydrocarbons with certified concentration around 4 ppb; absolute concentrations are propagated to a lower concentration standard in air that is used to calibrate the working standard. The GC-MS system is fully automated via the Linux-based chromatography software (GCWerks, gcwerks.com) adopted by the AGAGE (Advanced Global Atmospheric Gases Experiment) network. NMVOC dataset was categorized between "well-mixed" conditions (useful for deriving the long-term variability and tagged to the regional background) and "all-data" (also affected by regional emissions). The statistical procedure to identify the measurement periods representative of "well-mixed" conditions was presented by Lo Vullo et al. (2016a): more details are presented in the Chap. 2. Basically, the overall observed probability distribution function (PDF) of the dataset was decomposed into the sum of a Gaussian and a gamma distribution. The data falling within the Gaussian distribution corresponded to the observations carried out under "well-mixed" conditions, while the data following the gamma distribution were tagged to "non-well-mixed" conditions.

2.6 *Columnar NO_2*

Columnar nitrogen dioxide (NO_2) is measured at CMN since 1993 by zenith sky differential optical absorption spectroscopy (DOAS) by using a UV-Vis instrument of the GASCOD (gas analyzer spectrometer correlating optical differences) family spectrometers. This type of instruments has been developed and set up at the CNR-ISAC. The instrument used at CMN consists of a Cassegrain objective (150 cm, F/#5) as a receiving optics combined with 10 mm × 2 mm input slit which defines the instrumental field of view of 1.1E-5 sr. A specially designed holographic diffraction grating with 1200 l/mm provides sufficient spectral dispersion for detection of structured absorption features of the target gases. Spectral range available for exploration ranges from 250 nm to 900 nm detected by means of cooled CMOS diode array consisting of 512 pixels within single subinterval of 60 nm. In order to optimize CMOS exposure time, an autoranging procedure is built up within the instrument driving program ensuring enhanced S/N ratio. Ethernet link allows for the periodic downloading of the collected data. More details about inversion methodologies can be found in Petritoli et al. (2004).

During the years, maintenance procedures have been carried out in such a way as to guarantee unchanged performance of the instrument, assuring the continuity of the data series. DOAS measurements do not require absolute calibration. GASCOD instruments participated in two NDSC (Network for Detection of Stratospheric Change) intercomparison campaigns (Hofmann et al. 1995, Roscoe et al. 1999).

The measurements are performed twice a day during the sunrise and sunset periods (Fig. 2) within the solar zenith angle (SZA) range from 75 to 92°.

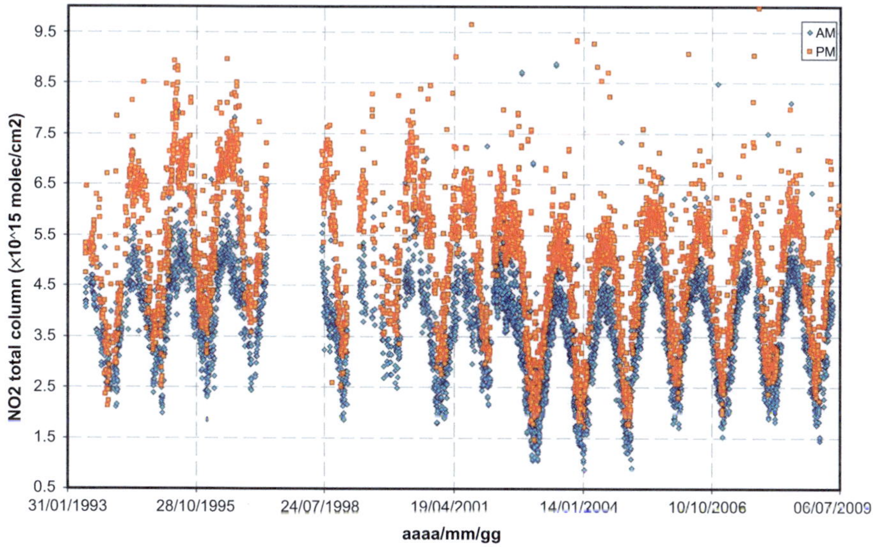

Fig. 2 Time plot of columnar NO_2 at CMN. Circles denote "sunrise" and squares "sunset" measurements

3 Overview of Variability: Typical Values and Seasonality

Over the period 1996–2015, surface O_3 showed an averaged mean value (calculated basing on daily mean values) of 53.5 ppb: a seasonal cycle with winter minima and spring-summer maxima (Fig. 1), something resulting in a broad spring-summer peak (e.g., during years 2003, 2005, 2014, and 2015), was evident at CMN (Fig. 1). Over the reporting period, the highest monthly mean value was observed in July (64.0 ± 0.7 ppb, ±95% confidence level) and the lowest in December (42.5.0 ± 0.5 ppb). Over 20 years of observations, the monthly peak was recorded in June for 7 years, in July for 6 years, in August for 3 years, and in May and April in 2 years. This behavior is typical for continental background sites (Tarasova et al. 2007), and the shape of the mean annual variation of O_3 at CMN is well comparable with the one observed at the other baseline sites in the Alps and Mediterranean basin (Cristofanelli et al. 2016). Indeed, the spring/summer maximum is considered to be a combination of the "hemispheric-scale" background and photochemical production from regional emissions (e.g., Monks et al. 2009).

CO (averaged value, 126.6 ppb) monthly average values were characterized by a peak in late winter-spring and a minimum in summer-autumn (Fig. 1). Over the period 2008–2015, the highest mean monthly value was observed in February (157.5 ± 5.1 ppb) and the lowest in November (113.0 ± 2.0 ppb). Since radical OH is the main atmospheric sink for CO, the highest values in late winter/spring can be explained by the longer CO lifetime in winter and subsequent accumulation, together with increased domestic emissions occurring at continental scale. The summer seasonal minimum is mainly driven by the seasonal maximum in OH

Table 3 Average NMVOC mixing ratios at CMN for all-data and well-mixed conditions (see Sect. 2.5)

Compound	CMN 2010–2014 Well-mixed	CMN 2010–2014 All-data	τ_{OH} (days)
Ethyne	115	185	14
Propane	291	406	11
Benzene	57	89	9.5
i-Butane	49	73	5.5
n-Butane	90	139	4.9
i-Pentane	49	84	3.2
n-Pentane	23	42	3.0
Toluene	33	109	2.1
Ethylbenzene	7	15	1.7
o-Xylene	5	19	0.9
m + p- Xylene	12	38	0.6

The atmospheric lifetimes with respect to OH reactivity are also reported, according to Lo Vollu et al. (2016a)

(Schultz et al. 2015). The late winter-spring maximum in a robust feature at CMN: for each individual year, the highest mean value was observed in February (4 years), March (3 years), and April (1 year). The average values and annual cycle of CO at CMN are in good agreement with those at other high-altitude mountain sites in Europe (Zellweger et al. 2009; Gilge et al. 2010).

NO (averaged value, 0.05 ppb) time series is also shown by Fig. 1. At CMN, NO seasonal cycle is characterized by peak values in February and March (0.13 ± 0.1 ppb). Very low values (often below instrumental detection limit) were observed from late spring to early autumn. This behavior reflect that observed for NO_2 (averaged value, 0.51 ppb), with average peak value in February (1.1 ± 0.2 ppb) and lower average values (ranging from 0.3 ppb to 0.6 ppb) in May to November. Even if caution should be deserved in commenting NO_2 at CMN before 2015 due to the use of a Mo converter in the CLD analyzer, these values fit nicely with those reported by Gilge et al. (2010) for the high mountain site Jungfraujoch (Switzerland 3480 m as.d.l.) and the rural site Hohenpeissenberg (Germany, 985 m a.s.l.).

At CMN, SO_2 showed an average value of 0.1 ppb with an average seasonal value characterized, as for other RGs, by a late winter/early spring maximum (mean values in February and March were 0.18 ± 0.08 ppb and 0.17 ± 0.04 ppb, respectively). The low SO_2 values observed at CMN testify that, due to the stringent emission reduction measures, SO_2 does not represent anymore a major concern for air pollution in Europe.

The average mean values of the 12 NMVOCs monitored at CMN are reported in Table 3. Averaging has been carried out for the entire dataset as well as for the "well-mixed" selection (see Sect. 2.5). As reported by Lo Vullo et al. (2016a), the NMVOC at CMN are in good agreement with those reported at remote sites in Europe. Over the period 2010–2014, the highest values were observed for propane, while the lowest concentrations were observed for the highly reactive ethylbenzene

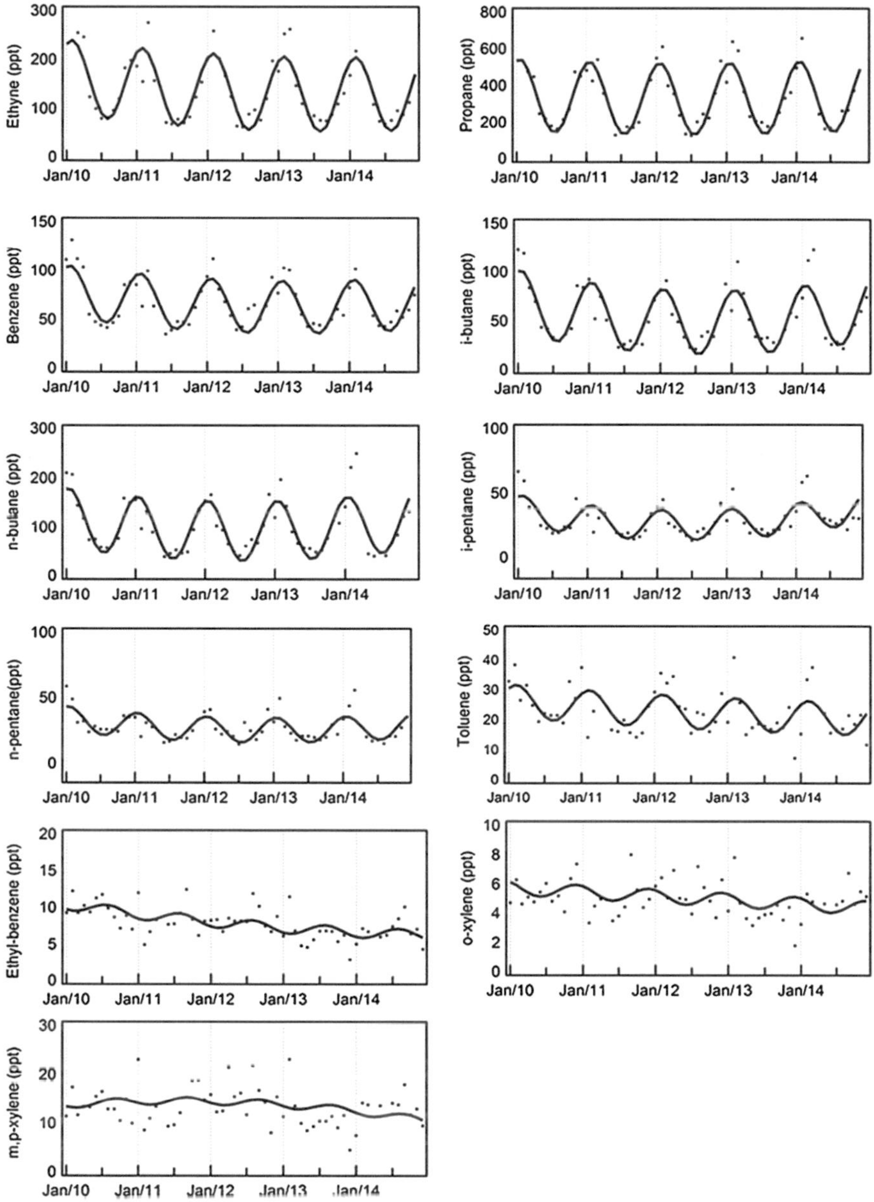

Fig. 3 Seasonal cycles of NMVOCs at CMN (2010–2014). Observed data (*dots*) and best fit (*solid line*)

and for the xylenes. Among the aromatic NMVOCs measured at CMN, benzene is the most abundant one, followed by toluene, ethylbenzene, and xylenes. The seasonal variability of NMVOCs under "well-mixed" conditions is reported by Fig. 3. The least reactive species (ethyne, propane, benzene, butanes, and pentanes) are characterized by maxima in cold months and minima in summer. More reactive species (toluene, ethylbenzene, and xylenes) show a much limited seasonal variability. These different behaviors were related to the different NMVOC reactivity with radical OH and the different contribution of continental versus regional sources. As an instance, the impact of European sources on the NMVOC mixing ratios is greater for the more reactive compounds, explaining the decrease in the seasonal variability with the increase in reactivity with OH radicals. Furthermore, aromatic compounds like toluene, ethylbenzene, and xylenes are subjected to higher sources release during the warmer months (Lo Vullo et al. 2016a).

4 Use of CMN Observations in Science Studies

4.1 Investigation of Vertical Air Mass Transport Processes

As being located at 2165 m a.s.l., CMN is considered representative of the background conditions of the free troposphere (see Chap. 1). However, different vertical air mass transports can affect the atmospheric composition variability at this measurement site. This is a matter of high scientific interest, since this represents a great opportunity to investigate the impact of emissions occurring at the Earth's surface to the free troposphere composition in the Mediterranean basin.

Surface O_3 variability and related atmospheric processes (i.e., vertical and horizontal air mass transport) have been extensively investigated at CMN since the begin of 2000s. Bonasoni et al. (2000a) provide a first multi-annual characterization of O_3 variability. By means of three-dimensional back-trajectories analysis, they defined "background conditions" when air masses reaching the measurement site did not penetrate below 780 hPa (i.e., the averaged CMN pressure level, assumed as an upper limit for the vertical extension of regional PBL) for at least 48 h before reaching the measurement site. This analysis showed that under "background conditions" the yearly maximum of O_3 is recorded in spring, while a secondary maximum appeared in summer. In contrast, if we consider "non-background conditions," the principal maximum was found in late summer. While the background spring maximum was considered representative of the seasonal O_3 cycle in the clean northern hemispheric, the late summer maximum traced the enhanced vertical transport of more polluted air masses from regional PBL (i.e., from Northern Italy and continental Europe) enriched by photochemical-produced O_3.

The role of thermal transport in enhancing the transport of polluted air masses from the lower troposphere/PBL up to the altitudes of the measurement site (which usually are considered representative of the free troposphere) is depicted by the average diurnal variations of O_3, CO, and specific humidity (a well-known tracer for air masses from the PBL, see Zellweger et al. 2000) in the different seasons (Fig. 4).

At CMN, clear diurnal cycles with higher SH values during the central part of the day (indicating systematic upward transport of air masses from PBL) were observed from spring to autumn (Fig. 4), while not statistically significant differences were

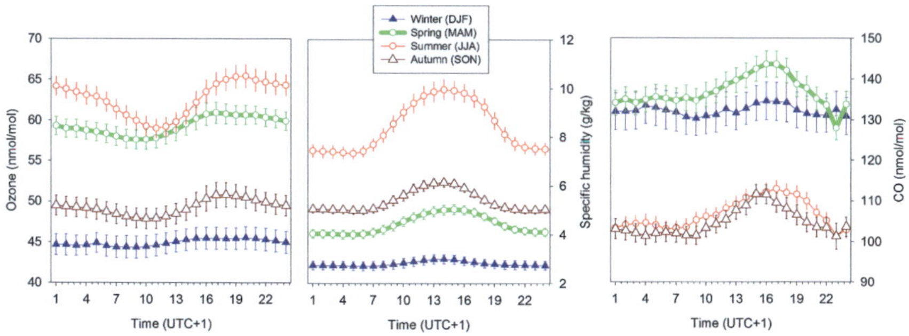

Fig. 4 Average seasonal diurnal O_3 mole fraction (*left*), specific humidity (*center*), and carbon monoxide (CO) at CMN (*vertical bars* denote the expanded uncertainty – $p < 0.05$ – of the mean)

observed for winter as a function of the time of day, thus indicating a limited influence of diurnal scale vertical mixing. Valley and upslope breezes developing during the central part of the day under high solar irradiance conditions favor the transport of humid and more polluted air masses from the PBL. Together with entrainment processes related to diurnal PBL growth, thermal wind systems can efficiently contribute to the export of pollutants to the free troposphere during daytime. As result, air masses with anthropogenic pollution can be systematically vented up to the measurement site during the central part of the day from spring to autumn as testified by the appearance of a diurnal CO peak. The average diurnal variation (mean daily maximum minus mean daily minimum) of O_3 is very small in winter and autumn (1–2 nmol/mol) and larger in spring (3 nmol/mol) and summer (6 nmol/mol). In particular, afternoon-evening O_3 observations in summer were significantly higher (at the 95% confidence level) in respect to midday hours, due to the diurnal scale vertical mixing of air masses richer in photochemically produced O_3 (Cristofanelli et al. 2009, 2015). As reported by Cristofanelli et al. (2007), these processes are particularly exacerbated during summer heat waves, when persistent dry and anticyclonic conditions enhance both the occurrence and intensity of photochemical smog episodes as well as the vertical transport of air masses from PBL to the free troposphere.

4.2 *Biomass Burning Emissions*

On a global scale, forest wildfires have a notable impact on the variability of tropospheric RGs in the Northern Hemisphere (e.g., Novelli et al. 2003). Significant increasing trend in fire activity was reported over Europe, Africa, and the USA (e.g., Westerling et al. 2006), and atmospheric compounds emitted by forest fires or produced by photochemical processes within smoke plumes can be transported over long distances, with possible impacts on air quality and climate of leeward regions.

At CMN, a typical event of long-range transport of open fire emissions was investigated by Cristofanelli et al. (2009) who analyzed the impact of a biomass burning (BB) plume originated from vegetation fires in the northern Africa. In particular, diffuse fire events affected the Atlas region from 25th to 30th August 2007. The advection of the BB products leads to significant increases of O_3 (+11%) and CO (+63%) in comparison to the average summer 2007 levels (64 ppb and 107 ppb, respectively). These evidences suggested that in the considered period, North Africa represented a significant source of RGs for the Southern Europe and Northern Italy. For attaining a better characterization of the BB products affecting CMN, the O_3 enhancement ratio (ER) was calculated. According to Andreae and Merlet (2001), the excess of an atmospheric species "X" normalized by the increase of CO can provide an estimation of the production by combustion processes of the species "X." CO can be considered the reference species because it represents one of the principal compounds emitted by combustion (Andreae and Merlet 2001) and has a long lifetime in the troposphere (Novelli 1999). The O_3/CO ERs (0.10–0.17 ppb/ppb) calculated by Cristofanelli et al. (2009) were within the range of those reported by previous investigations for relatively aged (2–4 days) BB plumes and suggested the occurrence of O_3 photochemical production during the transport to CMN.

Then, the impact of open fire emissions on O_3 and CO was investigated on a more systematic way by Cristofanelli et al. (2013). In particular, CMN observations from January 2009 to June 2012 have been analyzed and correlated with the outputs of the FLEXPART Lagrangian dispersion model (Stohl et al. 2005) which provided a model-calculated time series of CO mixing ratio (CO_{fire}) at CMN due to open vegetation fires (for more technical details see Cristofanelli et al. 2013). According to FLEXPART outputs, BB impact at CMN is maximized during the warm months (i.e., from July to September), but it had a significant contribution to RG variability only during specific transport events. In particular, to identify at CMN the presence of BB plumes, we considered the time periods for which FLEXPART CO_{fire} exceeded 10 ppb for more than 12 h. In this way, 16 events with a duration ranging from 12 to 72 h were identified, for a total of 21 days over 3 years. Based on the FLEXPART analyses, the transport of BB emissions occurred at very different scales: six events originated from the Southern Europe/Mediterranean basin, two events from the Balkan Peninsula, and one event from Russia. Five events were detected from North America, one from Central America, and one from Equatorial Africa. The majority of BB events (11) occurred from May to September, thus confirming that the transport of BB emissions to CMN is more likely during warm months.

We characterized in detail five "representative" events with respect to transport scales (i.e., global, regional, and local), source regions, and RG variations. We observed, for these events, very large variability of O_3/CO (from −0.22 to 0.71). The range of the positive O_3/CO ER (0.27–0.71) was in relatively good agreement with previous characterization of BB plumes. A negative value of −0.22 is observed during an event of fire plume from Equatorial Africa. Negative or not-significant O_3/CO ER was observed in aged BB plumes also by other studies (e.g., Val Martin et al. 2006; Real et al. 2007). This was attributed to the high aerosol concentrations within

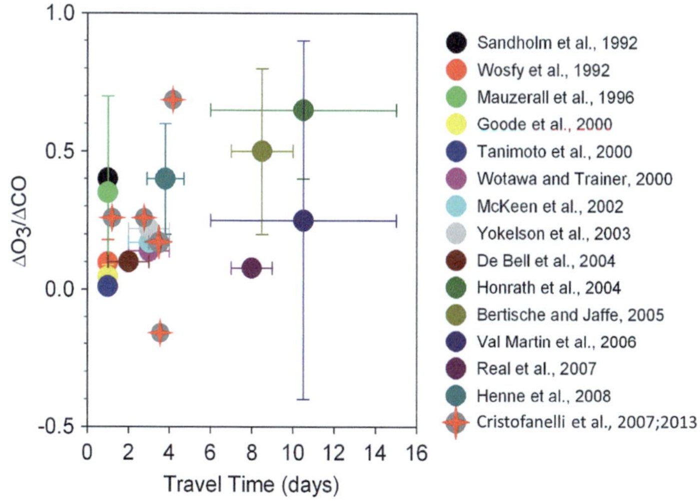

Fig. 5 ERs of O_3 (relative to CO) observed in previous investigations of BB plumes. *Vertical bars* denote the range of observed ERs. *Horizontal bars* denote the travel time for the observed BB plume (Modified by Cristofanelli et al. (2009), available at http://www.atmos-chem-phys.net/9/4603/2009/acp-9-4603-2009.html. Listed references can be found in Cristofanelli et al. 2009)

fire emission plumes (lowering down the O_3 formation efficiency due to an increased light extinction) or to other processes affecting nighttime chemistry (i.e., hydrolysis of N_2O_5) as well as O_3 loss by reaction with organic compounds during the long-range transport. In our case, a strong Saharan dust transport event accompanied the fire plume. Thus, the presence of mineral dust in the fire plume could possibly lead to heterogeneous removal of O_3 and its precursors (via $HNO_3$and NO_3 depletion) or/and to a decrease of photolysis rate (see Sect. 4.3).

The O_3 ER obtained at CMN was compared with that obtained at other measurement sites (Fig. 5). The different O_3/CO ER values were plotted as a function of the ages of the detected plumes (expressed as travel time). This analysis suggested that ER increases with the aging of plumes, indicating O_3 production within the BB plumes. This could be explained by the slow recycling of PAN, HNO_3, and organic nitrates, which favor the photochemical formation of O_3 in BB plumes by increasing the effective lifetime of NO_2.

4.3 *Impact of Mineral Dust Transport to Surface O_3*

Thanks to its location in the heart of the Mediterranean basin, CMN represents one of the first European mountain ridges that are affected by the northward transport of mineral dust from Africa. For this reason, it represents a very suitable site for investigating the influence of mineral dust transport to the background atmospheric composition of the Mediterranean basin. Indeed, Africa (in particular the Sahara and Sahel regions) is the largest source of mineral dust into the troposphere (Ginoux et al. 2012). The mobilized mineral aerosol can be transported over long distances by the synoptic circulation, with several direct and indirect effects on the atmospheric composition, air quality, and atmospheric radiative budget (Duchi et al. 2016 and references therein). In particular, mineral dust from northern Africa has two preferred transport regimes: over the Atlantic Ocean (Ben Ami et al. 2009) and over the Mediterranean basin (Israelevich et al. 2012 and references therein). Since dust particles present an efficient surface for chemical reactions and interact with solar radiation, several studies investigated its possible influence to the concentrations of atmospheric trace gases (e.g., Bauer et al. 2004; Fairlie et al. 2010; Ramachandra 2015). Indeed, as pointed out by laboratory and model studies (e.g., Bauer et al. 2004) as well as observations (Bonasoni et al. 2004), mineral dust may strongly affect the variability of tropospheric O_3. This point is particularly important for the Mediterranean region, which is characterized by high tropospheric O_3 levels (Zanis et al. 2014; Monks et al. 2009).

At CMN, mineral dust transport events were continuously identified since 2002. Coarse particle number concentration (i.e., aerosol particle with optical diameter $Dp > 1\ \mu m$) is continuously measured (time resolution, 1 min) by an optical particle sizer (OPC 1.108 Grimm) and used as "proxy" of mineral particles at the measurement site. Periods characterized by significant increases of coarse particle number concentrations with air masses originating from northern Africa (as deduced by means of air mass back trajectories or Eulerian simulation model, see Duchi et al. 2016 for more details) were recognized as affected by Saharan dust events.

In the framework of the MINATROC (Mineral Dust and Tropospheric Chemistry) EU Project, Bonasoni et al. (2004) provided a first systematic assessment of the impact of mineral dust transport O_3 variability at CMN. By analyzing 12 Saharan dust events, they found a strong anticorrelation between the volume of coarse particles and O_3. In particular, when southerly winds transported dust from northern Africa to CMN, surface O_3 showed a marked decrease (−5%) compared to typical free tropospheric conditions. For a dust intrusion event which occurred at CMN during the MINATROC intensive field campaign (from 1st June to 7th July 2000), Hanke et al. (2003) found out that the mixing ratio of HNO_3 started to continuously decrease at the onset of the dust event, until the point it was "virtually" absent at the end of the event. These measurements provided for the first time a strong observational indication of an efficient uptake of gaseous HNO_3 by mineral-dust aerosol particles, thus implying that mineral dust represents a significant sink for HNO_3 in the troposphere, with implications on the photochemical cycles and O_3 production.

More recently, Duchi et al. (2016) performed a long-term (2002–2015) investigation of dust transport events (DTEs) from the Sahara desert and their impact on surface O_3 variability at CMN. In agreement with the earlier investigation by Bonasoni et al. (2004), O_3 significantly decreased when mineral dust increased. This negative correlation was particular evident during spring and summer, when the maximum O_3 decrease was found to be −4.1 ppb (−7%) in respect to "typical" values obtained by applying a 21-day running mean to hourly averages. However, as shown by concomitant CO measurements in the period 2007–2011, the mixing with polluted air masses rich in combustion emissions could limit the O_3 depletion observed within dusty air masses. Indeed, by neglecting DTEs affected by mixing with polluted air masses, as deduced by the increase of CO mixing ratio, the maximum O_3 decrease was −8% during spring and higher than −9% during summer. To rule out the possibility that the low O_3 mixing ratio observed during dust events at CMN could be only related to the origin or to the passage of air masses in the O_3-depleted Saharan PBL, Duchi et al. (2016) calculated the variability of O_3 at CMN during air mass transport from northern Africa but without any significant presence of coarse particles (i.e., mineral dust). The result (−0.1 ± 0.1 ppb) indicated that at CMN the passage through the Saharan PBL is not an explanation of the low O_3 observed during DTEs. Since CMN is well representative of the central Mediterranean baseline conditions, these results represent an important contribution for a better understanding of the impact of natural processes (like mineral dust emission and transport) to the variability of this well-known SLCF and pollutant.

4.4 Deep Stratospheric Intrusions

Stratosphere-troposphere exchange (STE) is a driving process for chemical composition variability of both within the stratosphere and into the troposphere (Holton et al. 1995; Stevenson et al. 2006). As an instance, nowadays, there is a large debate about the possibility that long-term changes in the stratosphere-to-troposphere transport could affect long-term tropospheric O_3 trends (Cooper et al. 2014). A specific aspect of STE are the stratospheric intrusions (SIs), which are defined as the downward transport of stratospheric air masses relatively deep into the troposphere. Several mechanisms promote SIs: tropopause folds, cutoff lows, upper-level fronts, high-pressure systems, subtropical jet streams and streaks, and potential vorticity (PV) streamers (e.g., Appenzeller and Davies 1992; Holton et al. 1995; Stohl et al. 2003; Sprenger et al. 2007). As pointed out by Zanis et al. (1999), the downward motion of stratospheric air in the lower troposphere can occur in a direct or indirect way. While in the case of direct SI, stratospheric air largely maintains its original properties as it reaches the lower troposphere by rapid vertical transport, indirect SI reaches the lower troposphere after a sequence of transport "steps," providing the opportunity to experience mixing with tropospheric air. In these cases, because of mixing processes acting on small and intermediate (100 km) scales, stratospheric air masses quickly lose their original properties (Appenzeller and Davies 1992), making difficult the identification of stratospheric air masses at low altitudes. Therefore, high mountain sites are privileged platform for SI observations, but the capability to detect SI relies on the continuous monitoring of specific atmospheric tracers. Different experimental tracers can be used to identify SIs by in situ measurements: relative humidity, ^{7}Be, ^{10}Be, O_3, and atmospheric pressure variability (see also Chap. 4). These observations are often integrated with datasets of vertical profiling (radio/ozonesondes or lidar), satellite observations (e.g., total column of ozone), and various kinds of numerical model outputs. At CMN, SI events have been investigated both on single event basis (e.g., Bonasoni et al. 2000b) and by multi-year approach (e.g., Cristofanelli et al. 2006; Putero et al. 2016). During two case studies that occurred on March and May 1996, Bonasoni et al. (2000b) reported significant O_3 increases (up to 100 ppb on 1-min average) during the occurrence of SI. Cristofanelli et al. (2006) investigated the impact of SIs on O_3 variability from 1998 to 2003. The occurrence of SI events at CMN was corroborated by the analyses of in situ stratospheric tracers (^{7}Be and relative humidity), together with potential vorticity[1] in air masses arriving at the measurement site (as deduced by back-trajectories calculation) and tropopause height over Northern Italy (as obtained by radio-soundings). Different selection criteria based on the combination of these observed and modeled stratospheric tracers were adopted to identify SIs. Basing on this study, on average, SI affected CMN for 36 days/year (i.e., 10% of the period), with a clear winter maximum (24% of days in January) and a spring-summer minimum (4–5% in April and August). The contribution to O_3 was largest in winter. For "direct" SI, the O_3 increase by 8% (ranging from 5% in spring to 9% in winter) with respect to the monthly running mean. An increase of 3% was found for "indirect" SI (ranging from 4% in winter to less than 1% in summer). The low O_3 increase observed during summer probably stressed the importance of photochemical O_3 production over Mediterranean basin during this season in respect to the stratospheric contribution.

More recently, Putero et al. (2016) investigated the long-term variability of SI at CMN by using STEFLUX (stratosphere-to-troposphere exchange flux). This is a Lagrangian tool which relies on the stratosphere-to-troposphere exchange climatology compiled at ETHZ by Škerlak et al. (2014), which makes use of the ERA-Interim reanalysis dataset from the ECMWF. STEFLUX detects the air parcels originating in the stratosphere and entering a tropospheric 3-D target box during a specific time window: this time period is flagged as affected by SI. For CMN it was applied to the period 1979–2015. In nice agreement with the earlier work by Cristofanelli et al. (2006), it was found out that 9% of the period was affected by SI (Fig. 6) with an average event duration of 2.1 days. STEFLUX results confirmed that the annual cycle of SI frequency is maximized in winter and minimized in summer at CMN. No significant trends in the SI frequency were discernible: only a weak positive tendency of 0.03% yr. was detected. Interestingly, a correlation was found between SI frequency and ENSO (El Nino Southern Oscillation). This would suggest that possible changes in the frequency or intensity of ENSO due to climate change can affect SI occurrence in the Mediterranean basin.

[1] Potential vorticity is defined as $PV = -g\frac{\delta\theta}{\delta p}\left(\xi_\theta + f\right)$ where g denotes the gravity acceleration, θ the potential temperature, ξ_θ is the component of the curl of wind vector normal to an isentropic surface, and f is the Coriolis parameter.

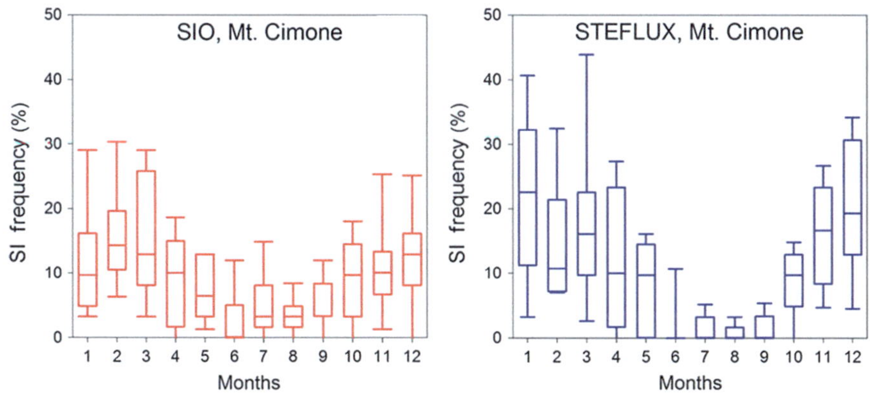

Fig. 6 *Box-and-whisker* plot of the annual variation in SI frequency at CMN as deduced by experimental data (*left*) and STEFLUX (*right*) (Modified by Putero et al. (2016) available at http://www.atmos-chem-phys.net/16/14203/2016/acp-16-14203-2016.html)

4.5 Transport of Anthropogenic Pollution

Several investigations pointed out the impact of transport of anthropogenic pollution to RG variability at CMN. During the PEGASOS (Pan-European Gas-AeroSOls Climate Interaction Study) experiment, carried out on summer 2012, RG (O_3, CO, NO_x) variability was investigated and the influence of anthropogenic emission sources estimated. In particular, Cristofanelli et al. (2016) analyzed the variability of O_3/NO_x and CO/NO_x ratios to qualitatively evaluate the proximity to major emission sources and photochemical processing (Morgan et al. 2010; Parrish et al. 2009): $O_3/NO_x \leq 50$ tagged near outflow emissions (N-OUT), $50 < O_3/NO_x \leq 90$ far outflow emissions (F-OUT), while $O_3/NO_x > 90$ characterized background conditions (BKG). During the analyzed period, 38% and 39% of observations were tagged to the far outflow and background regimes, indicating that also under typical summer conditions CMN is well representative of the background conditions of the Mediterranean basin.

The analysis of RGs as a function of emission source proximity (as deduced by NO_x/CO ratio) pointed out that all the atmospheric compounds showed decreasing values shifting from N-OUT to BKG with statistical significant (at the 95% confidence level) differences among the three regimes (Table 4). The highest decrease was observed for NO_x (−74%), while the lowest decrease was observed for O_3 (−4.7%). With the purpose of taking into account the diurnal variability which affected RGs at CMN, we classified RGs also as a function of time periods influenced by thermal upward transport (07–21 UTC + 1, hereinafter "daytime") or more representative of the free troposphere (23 – 06 UTC + 1, "nighttime"). "Nighttime" data were only marginally affected by "near" emissions (only 5% of data), and NO_x/CO values representative of BKG conditions were dominant (69.6% of data). The decreasing trend of RG values from N-OUT to BKG was evident also for

Table 4 Average mean value (Avg) and number of data (ND) for O_3 (ppb), CO (ppb), NOx (ppb as a function of the different photochemical processing (near outflow, N-OUT; far outflow, F-OUT; background, BKG)

		N-OUT		F-OUT		BKG	
Data selection	RG	Avg	ND	Avg	ND	Avg	ND
All data	O_3	62.5 ± 1.7	120	57.9 ± 1.2	232	59.1 ± 1.1	234
	CO	131.7 ± 2.5	116	111.9 ± 1.9	209	99.6 ± 1.6	188
	NO_x	1.62 ± 0.08	131	0.88 ± 0.03	228	0.42 ± 0.02	193
9-21 UTC + 1	O_3	63.2 ± 1.7	110	56.8 ± 1.2	157	57.0 ± 1.4	98
	CO	133.4 ± 2.3	108	114.3 ± 1.9	145	102.2 ± 2.2	84
	NO_x	1.64 ± 0.08	121	0.86 ± 0.04	166	0.42 ± 0.02	91
23-6 UTC + 1	O_3	55.5 ± 7.5	8	59.5 ± 2.6	57	60.6 ± 1.8	108
	CO	109.3 ± 8.1	7	103.9 ± 5.1	47	96.9 ± 2.6	80
	NO_x	1.46 ± 0.23	8	0.89 ± 0.06	44	0.42 ± 0.03	81

Upper, all data; middle, "daytime"(9-21 UTC + 1) data; bottom, "nighttime" data (23-6 UTC + 1). Adapted by Cristofanelli et al. (2015)

"nighttime" and "daytime" selections. Only for "nighttime" O_3, an increasing tendency was observed. This probably indicated the effect of NO titration in air masses more affected by fresher anthropogenic emissions (N-OUT).

The analysis of O_3 and CO average values as a function of wind direction and O_3/NO_x regimes showed that for "near" emissions (N-OUT regime) air masses from NE (i.e., from the Po basin) and SW (i.e., from Tyrrhenian Sea) sectors controlled the overall mean of CO and O_3. In particular, for O_3 the highest values were related to NE winds, indicating the role of the Po basin as important source of photochemically produced O_3. This is a point of high concern since some of these pollutants (i.e., O_3 and eqBC) are known to be short-lived climate forcers (UNEP and WMO 2011) and, once in the free troposphere, can significantly affect regional climate. For the observation period affected by "far" sources or background conditions (i.e., $O_3/NO_x > 50$), the wind sectors from SW dominated the O_3 and CO variability of atmospheric tracers/pollutants. The results obtained during the PEGASOS experiment leaded to the conclusion that: (1) diurnal thermal transport of air masses from the Po basin was associated to the highest RG values, and (2) even during summer season, when direct transport of anthropogenic pollutants from PBL is maximized at CMN, this measurement site is affected for the most part of the time (79%) by aged anthropogenic emissions or background conditions.

With the aim of disentangle the different contributions to O_3 and CO among regional and long-range pollution, Cristofanelli et al. (2013) analyzed the outputs of FLEXPART dispersion model at CMN for year 2009–2012. In particular, FLEXPART provided a time series for "anthropogenic" emitted CO at CMN (CO_{ant}), basing on the EDGAR version 3.2 emission inventory for the year 2000 outside North America and Europe, EMEP emission inventory for the year 2005 for Europe and the emission inventory by Frost et al. (2006) for North America. As suggested by FLEXPART simulations, 20-day-old anthropogenic CO contributions can explain up to 17.4% of the mean CO value observed at CMN. In particular, the months from April to September appeared to be strongly affected by transport of anthropogenic pollution. Indeed, this contribution explained from 22% (in January–April) to 37% (in May–September) of observed variability for CO and from 1% (in October–December) to 19% (in May–September) for O_3. Thanks to FLEXPART use, it was possible to investigate, as a function of emission ages, the geographical origin of the CO_{ant} at CMN. From this analysis, it was find out that emissions located inside Europe are predominant for emission ages lower than 120 h, while emissions located in North America, Asia, and Africa were dominant for emission ages of CO_{ant} larger than 120 h.

In agreement with the results from the PEGASOS experiment in summer 2012, during May–September 2009–2012, a significant positive correlation (*R*, 0.79) was found between CO and O_3. The highest CO and O_3 values were observed together with high value of equivalent black carbon (>400 ng m^{-3}), a well-known tracer for combustion processes. This clearly underpins the role of regional pollution (from PBL of Northern Italy and Continental Europe) in affecting summer RG variability at CMN. During cold months (October–December), CO and O_3 were negatively correlated (*R*, −0.51) probably due to different processes interacting each other (i.e., transport of air masses from upper troposphere/lower stratosphere intrusions, O_3 titration by NO in the polluted European PBL).

The impact of sources and transport to anthropogenic on NMVOCs has been investigated by Lo Vullo et al. (2016b) for the period 2010–2014. By means of a multifactor principal component analysis (PCA), different source categories have been identified for 12 NMVOCs (Table 3). During winter, two main source categories for NMVOCs were identified: aged air masses related to long-range transport of vehicular exhausts (WF1) and relatively fresh emission of industrial solvent evaporation (WF2). These two components accounted for 78% and 12% of the observed NMVOC variability during winter. During summer, three main source components were identified: aged air masses due to long-range transport (SF1, accounting for 62% of NMVOC variability), emissions from fuel evaporation which experienced medium-range transport (SF2, 16%), and local sources of solvents (SF3, 8%).

4.6 Long-Term Trends

At CMN, long-term trends were assessed for O_3 and anthropogenic NMVOC. Cristofanelli et al. (2015) analyzed an "extended" O_3 data series by considering earlier measurements carried out from 1991 to 1993. For the overall time interval 1991–2011, significant (95% confidence level) positive linear trends (calculated as linear fitting of de-trended data) have been obtained for monthly average values as well as for the 25th and 75th percentiles. For the shorter period 1996–2011, no significant O_3 trends were detected at CMN, in fairly good agreement with observations at other high mountain stations in Europe (e.g., Gilge et al. 2010; Parrish et al. 2012) and in the Mediterranean basin (Kleanthous et al. 2014; Saliba et al. 2008). By considering the 1991–2011 period, we detected a slowing down of the O_3 growth rates in all seasons at CMN and a lower mean O_3 mole fraction observed during recent years (i.e., from 2009 to 2011, 51.9 ± 8.3 nmol/mol), in agreement with other Alpine and Mediterranean baseline sites. In particular, as suggested by Parrish et al. (2012), to detect possible changes in the growth rates with time, we analyzed the seasonal long-term variability by interpolating the annual mean O_3 with a polynomial quadratic fit. Statistically significant negative values of the second-order parameter ($d[O_3]/dt^2$) are found at CMN in all seasons except autumn, indicating a slowing of O_3 increase rate. For summer, a significant negative "acceleration" is detectable also for the period 1996–2011. This would reflect the decrease of O_3 precursor emissions within Europe which led to reductions in peak summertime O_3 also at surface continental sites (The Royal Society 2008). This would also affect CMN observations since, especially during summer, the northern Apennines can be exposed to polluted air masses from the regional and continental PBL (see Cristofanelli et al. 2007; Cristofanelli et al. 2013).

With the aim of investigating the long-term O_3 trend over the period 1996–2013, we adopted the Theil-Sen method. The basic idea is as follows. Given a set of n x, y pairs, the slopes between all pairs of points are calculated. The Theil-Sen estimate of the slope is the median of all these slopes. The advantage of using the Theil-Sen estimator is it is resistant to outliers (Carslaw and Ropkins 2012). The Theil-Sen method was applied to the deseasonalized monthly values of O_3 (Fig. 7). For CMN, over the whole measurement period, a negative trend (significant at the 99% confidence level) of -0.18 ± 0.07 ppb year^{-1} was obtained (see also Appendix 1).

NMVOC trends were evaluated by Lo Vullo et al. (2016a) over the period 2010–2014. Lo Vullo et al. (2016a) estimated the trend values by using the nonparametric Mann-Kendall test on the whole dataset as well as on the data selection for "well-mixed" conditions (see Sect. 2.5). Over the period 2010–2014, ethyne showed a decreasing trend of −2.7%/year, which is consistent with abatement measures on the exhaust emissions from motor vehicles. No significant trends are observed for "well-mixed" propane and benzene. Propane is emitted mainly from natural gas leakages, while benzene is a component of unburned fossil fuels and a by-product of the incomplete combustion of other hydrocarbons in fossil fuels. European countries adopted stringent regulations on the use of benzene. The absence of statistically significant trends at CMN is in agreement with the results by Grant et al. (2011) at the remote station of Mace Head (Ireland).

For the "all-data" selection, butanes and pentanes showed a decrease at CMN, in good agreement with observations at European rural and urban sites due to the reduction of evaporative emissions from petrol. For "well-mixed" conditions, no significant decrease was reported for butanes and i-pentane, while n-pentane decreased by −5.7%/year. This would indicate that n-pentane at CMN is affected by emission sources different from petrol evaporation. For toluene, used as feedstock and solvent, the whole dataset from CMN shows a decrease of −3.6% per year, reflecting the effect of the EU VOC Solvent Emission Directive (1999/13/CE European Directive with amending) setting the limitation of NMVOCs used in the formulation of organic solvents. "Well-mixed" ethylbenzene, used as solvents in paints and coating but also contained in vehicle exhausts, showed a significant decrease of 8.7% per year. No significant variations were observed for xylenes over the investigated period.

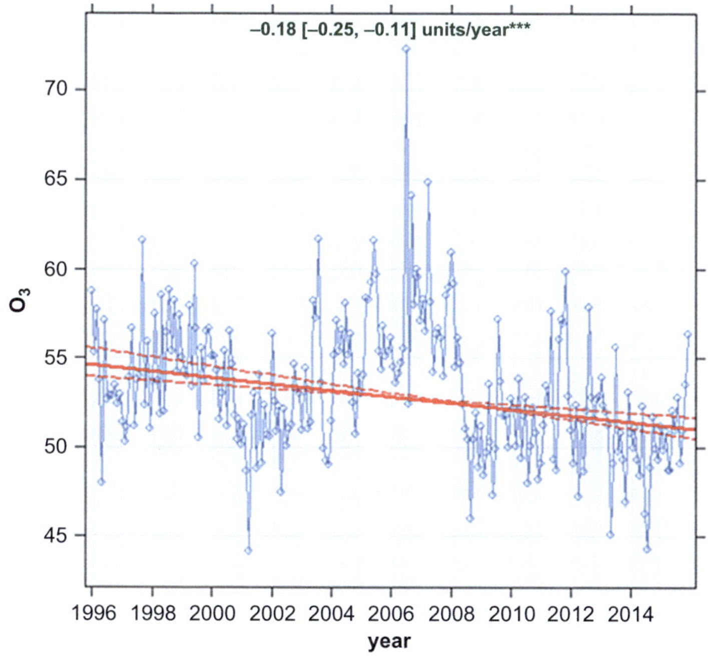

Fig. 7 Time series of deseasonalized monthly O_3 values at CMN (expressed as ppb). The results of the trend calculated by the Theil-Sen method are reported by the *bold line* together with the 95% confidence interval (*dotted lines*). The annual trend value is reported over the plot

Kostadinov et al. (2011) investigated the long-term variations of tropospheric NO_2 obtained at CMN from 1993 to 2009 as a function of solar zenith angles (90° and 88°) and time period (sunset and sunrise). By the implementation of different model approach and tests, negative NO_2 trends were detected for measurements both at sunset (ranging from -11 ± 2 to $-10 \pm 2\%$ per decade) and sunrise (ranging from -10 ± 3 to $-8 \pm 2\%$ per decade). These results appeared to be in agreement with trend analysis (Gruzdev 2009) at other Northern Hemisphere stations. Kostadinov et al. (2011) attributed this negative tendency to the interplay between chemical and physical atmospheric processes influenced by both natural and anthropogenic impacts.

4.7 *Specific Applications*

4.7.1 Model Validation

Atmospheric observations of RG variability at a mountain peak-like CMN are useful to verify the performance of atmospheric models (both climate and chemistry-transport models) in reproducing specific processes.

Together with observations at other four high mountain stations in Europe (Jungfraujoch-CH, Sonnblick-A, and Zugspitze-D), in the framework of the EU Project STACCATO (Stratosphere-Troposphere Exchange in a Changing Climate on Atmospheric Transport and Oxidation Capacity), CMN data have been used to verify the ability of seven different models in reproducing SI events over Europe (Cristofanelli et al. 2003). Measurements of surface O_3, ^{7}Be, ^{10}Be, and relative humidity were used to characterize the developments of a deep SI event which occurred on May 1996 (Bonasoni et al. 2000b) and to compare the observational evidences with the model output. It was found that only the models with sufficient horizontal spatial resolution (i.e., LAGRANTO, FLEXTRA, and FLEXPART) were able to completely capture the development of the stratospheric air intrusion down to CMN altitude.

More recently a system for the near-real-time data delivery was implemented at CMN to contribute to the online data verification of the MACC (Monitoring Atmospheric Composition and Climate) Global Near-Real-Time model production run for RGs and aerosol. MACC represented the preoperational version of Copernicus Atmosphere Monitoring Service (CAMS, see http://www.copernicus.eu/) EU Program. It produced daily analyses and 5-day forecasts in near-real-time and provided the boundary conditions for the MACC regional air quality products (RAQ) and other downstream users. It was the only system worldwide that is run in near-real-time to produce global reactive gases and aerosol fields. To this aim, starting from 2009, hourly average values of O_3, CO, NO, NO_2, and SO_2 have been delivered on a daily basis to DWD (Deutscher Wetterdienst) from CMN. Before delivery to DWD, CMN data are subjected to basic quality check controls including data plausibility checks, maximum internal variability check, and internal consistency checks. Together with observations from other WMO/GAW stations and EMEP measurement sites, CMN data have been extensively used in Wagner et al. (2015) to investigate the MACC model's ability to simulate tropospheric mixing ratios of surface O_3 and CO in the period between 2009 and 2012.

Currently, these data are routinely used by CAMS (https://atmosphere.copernicus.eu/) for producing quarterly validation reports of the near-real-time global atmospheric composition service. These reports are available at the web address https://atmosphere.copernicus.eu/quarterly_validation_reports, while online comparison between CMN observations and model products can be found at http://macc.copernicus-atmosphere.eu/d/services/gac/verif/grg/gaw/.

4.7.2 Ground-Truthing

Tropospheric column measurements of NO_2 at CMN were used to assess the validity of spaceborne measurements. Indeed, new methods developed to obtain tropospheric column from satellite platforms need to be compared with independent measurements to understand the nature of the information that can be used for scientific studies and operational purposes. In particular, Petritoli et al. (2004) compared the tropospheric NO_2 measurements at CMN by the GASCOD instrument with GOME (Global Ozone Monitoring Experiment) observations. This study pointed out that the mixing properties of the atmosphere appeared to be the most important parameter for a valid comparison of the measurements. It was concluded that GOME NO_2 tropospheric column measurements could be used for NO_2 monitoring only with the support of auxiliary measurements or model calculations.

Moreover, Petritoli et al. (2003) performed a preliminary comparison of tropospheric NO_2 data at CMN with the total column measurements performed by SCIAMACHY spectrometer on board ENVISAT platform. Also in this case a particular attention was devoted to define a correct comparison method. Indeed, besides having a different geometrical approach, CMN measurements were performed at sunrise and sunset while the ENVISAT satellite overpass the measurement site between 10 and 11 AM local time. At the time of this study, the availability of SCIAMACHY NO_2 total column from the data processor SCIA/3.53 version was restricted to part of October and November 2002. Thus, also taking into account the even smaller ensemble of overpasses over CMN, only a few days of concurrent measurements were available for the comparison. Even if the obtained results only provided a preliminary indication on the real quality of SCIA/3.53 version of data processing, a tendency toward an overestimation (by 20%) of the NO_2 total column in respect to CMN data was assessed.

5 Conclusions

High-quality atmospheric measurements of several reactive gases (O_3, CO, NO_x, SO_2, NMVOC) carried out at CMN in the last decades, allowed to disentangle the contribution of different processes (transport of anthropogenic pollution, biomass burning emissions, stratospheric air mass intrusions, mineral dust transport) in determining their variability in the lower troposphere of the Mediterranean basin. CMN represents the measurement site in the Mediterranean region, with the longest period of simultaneous and continuous observations of these compounds. Data quality protocols are implemented in the framework of WMO/GAW and ACTRIS/ACTRIS-2 Projects, thus allowing the intercomparability of CMN observations with other international laboratories.

A number of atmospheric processes able to affect RG variability in the Mediterranean basin have been investigated at CMN along the years: vertical and horizontal air mass transport, stratospheric intrusions, biomass burning, dust outbreaks, and transport of anthropogenic pollution. The information collected at CMN were used to evaluate the long-term variability (i.e., trends) for some of these molecules (O_3 and NMVOC) as well as to validate atmospheric models. These research activities represent a notable contribution also for evaluating the effectiveness of measures for reducing the anthropogenic emissions of RGs. For these reasons, the RG data are submitted year by year to the Global Atmosphere Program by WMO, while near-real-time data delivery services have been activated toward CAMS/COPERNICUS.

Acknowledgments The results described in this chapter have been obtained thanks to the activities carried out in the framework of different research projects funded by the European Commission and Italian Ministry of Education, University, and Research, the Italian Space Agency and ESA: VOTALP (Contract: ENV4-CT95-0025 and IC20-CT96-0015), VOTALP II (ENV4-CT97-0413 and IC20-CT97-0039), STACCATO (EVK2-CT-1999-00050), MINATROC (EVK2-CT-1999-00003) ACTRIS-2 (Aerosols, Clouds, and Trace gases Research InfraStructure) under the European Commission Horizon 2020 Research and Innovation Framework Programme (H2020-INFRAIA-2014-2015), QUITSAT (Qualità dell'aria mediante l'Integrazione di misure da Terra, da Satellite e di modellistica chimica multifase e di Trasporto, ASI contract I/035/06/0); DUPPOLPO (ESA Contract No. 15564/01/I-LG), EU-QUILT (EVK2-2000-00545), SHARE (Stations at High Altitude for Research on the Environment), and the Project of National Interest NEXTDATA.

References

Andreae MO, Merlet P (2001) Emission of trace gases and aerosols from biomass burning. Glob Biogeochem Cy 15:955–966

Appenzeller C, Davies HC (1992) Structure of stratospheric intrusions into the troposphere. Nature 358:570–572

Bauer SE, Balkanski Y, Schulz M et al (2004) Global modelling of heterogeneous chemistry on mineral aerosol surfaces: the influence on tropospheric ozone chemistry and comparison to observations. J Geophys Res 109. doi:10.1029/2003JD003868

Ben-Ami Y, Koren I, Altaratz O (2009) Patterns of North African dust transport over the Atlantic: winter vs. summer, based on CALIPSO first year data. Atmos Chem Phys 9:7867–7875

Bonasoni P, Evangelisti F, Bonafè U et al (2000a) Stratospheric ozone intrusion episodes recorded at Mt.Cimone during the VOTALP project: case studies. Atmos Environ 34(9):1355–1365

Bonasoni P, Stohl A, Cristofanelli P et al (2000b) Background ozone variations at Mt. Cimone Station. Atmos Environ 34:5183–5189

Bonasoni P, Cristofanelli P, Calzolari F et al (2004) Aerosol-ozone correlations during dust transport episodes. Atmos Chem Phys 4:1201–1215

Carslaw DC, Ropkins K (2012) Openair – an R package for air quality data analysis. Environ Model Soft 27-28:52–61

Collins WJ, Derwent RG, Garnier B et al (2003) Effect of stratosphere-troposphere exchange on the future tropospheric ozone trend. J Geophys Res 108(D12):8528

Cooper OR, Parrish DD, Ziemke J et al (2014) Global distribution and trends of tropospheric ozone: an observation-based review. Elem Sci Anth. doi:10.12952/journal.elementa.000029

Cristofanelli P, Bonasoni P, Collins W, Feichter J, Forster C, James P, Kentarchos A, Kubik PW, Land C, Meloen J, Roelofs GJ, Siegmund P, Sprenger M, Schnabel C, Stohl A, Tobler L, Tositti L, Trickl T, Zanis P (2003) Stratosphere-to-troposphere transport: A model and method evaluation. J Geophys Res Atmos 108(12):STA 10-1–STA 10-23

Cristofanelli P, Bonasoni P, Tositti L et al (2006) A 6-year analysis of stratospheric intrusions and their influence on ozone at Mt. Cimone (2165 m above sea level). J Geophys Res 111:D03306

Cristofanelli P, Bonasoni P, Carboni G et al (2007) Anomalous high ozone concentrations recorded at a high mountain station in Italy in summer 2003. Atmos Environ 41:1383–1394

Cristofanelli P, Marinoni A, Arduini J et al (2009) Significant variations of trace gas composition and aerosol properties at Mt. Cimone during air mass transport from North Africa – contributions from wildfire emissions and mineral dust. Atmos Chem Phys 9:4603–4619

Cristofanelli P, Fierli F, Marinoni A et al (2013) Influence of biomass burning and anthropogenic emissions on ozone, carbon monoxide and black carbon at the Mt. Cimone GAW-WMO global station (Italy, 2165 m a.s.l.) Atmos Chem Phys 13:15–30

Cristofanelli P, Scheel H-E, Steinbacher M et al (2015) Long-term surface ozone variability at Mt. Cimone WMO/GAW global station (2165 m a.s.l., Italy). Atmos Environ 101:23–33

Cristofanelli P, Landi TC, Calzolari F et al (2016) Summer atmospheric composition over the Mediterranean basin: investigation on transport processes and pollutant export to the free troposphere by observations at the WMO/GAW Mt. Cimone global station (Italy, 2165 m a.s.l.) Atmos Environ 141:139–152

Duchi R, Cristofanelli P, Landi TC et al. (2016) Long-term (2002–2012) investigation of Saharan dust transport events at Mt. Cimone GAW global station, Italy (2165 m a.s.l.). Elem Sci Anth. doi:10.12952/journal.elementa.000085

Fairlie TD, Jacob DJ, Dibb JE et al (2010) Impact of mineral dust on nitrate, sulfate, and ozone in transpacific Asian pollution plumes. Atmos Chem Phys 10:3999–4012

Frost GJ, McKeen SA, Trainer M et al (2006) Effects of changing power plant NOx emissions on ozone in the eastern United States: proof of concept. J Geophys Res. doi:10.1029/2005JD006354

Gilge S, Plass-Duelmer C, Fricke W, Kaiser A, Ries L, Buchmann B, Steinbacher M (2010) Ozone, carbon monoxide and nitrogen oxides time series at four alpine GAW mountain stations in central Europe. Atmos Chem Phys 10:12295–12316. doi:10.5194/acp-10-12295-2010

Ginoux P, Prospero JM, Gill TE et al (2012) Global-scale attribution of anthropogenic and natural dust sources and their emission rates based on MODIS deep blue aerosol products. Rev Geophys. doi:10.1029/2012RG000388

Grant A, Yates EL, Simmonds PG et al (2011) A 5 year record of high-frequency in situ measurements of non-methane hydrocarbons at Mace head, Ireland. Atmos Meas Tech Discuss 4:913–937

Gruzdev AN (2009) Latitudinal structure of variations and trends in stratospheric NO2 result. Int J Rem Sens 30:4227–4246

Hanke M, Umann B, Uecker J et al (2003) Atmospheric measurements of gas-phase HNO3 and SO2 using chemical ionization mass spectrometry during the MINATROC field campaign 2000 on Monte Cimone. Atmos Chem Phys 2003. doi:10.5194/acp-3-417-2003

Henne S, Brunner D, Folini D (2010) Assessment of parameters describing representativeness of air quality in-situ measurement sites. Atmos Chem Phys 2010. doi:10.5194/acp-10-3561-2010

Hofmann D, Bonasoni P, de Maziere M et al (1995) Intercomparison of UV/visible spectrometers for measurements of stratospheric NO2 for the network for the detection of stratospheric changes. J Geophys Res 100:16765–16791

Holton JR, Haynes PH, McIntyre ME et al (1995) Stratosphere-troposphere exchange. Rev Geophys 33:403–439

Israelevich PL, Ganor E, Alpert P et al (2012) Predominant transport paths of Saharan dust over the Mediterranean Sea to Europe. J Geophys Res. doi:10.1029/2011JD016482

Kleanthous S, Vrekoussis M, Mihalopoulos N et al (2014) On the temporal and spatial variation of ozone in Cyprus. Sci Tot Environ 476–477:677–687

Kostadinov I, Petritoli A, Giovanelli G et al (2011) Stratospheric NO2 trends over the high mountain 'Ottavio Vittori' station, Italy. Int J Rem Sens. doi:10.1080/01431161.2010.517799

Lo Vullo EL, Furlani F, Arduini J et al (2016a) Non-methane volatile organic compounds in the background atmospheres of a southern European mountain site (Mt. Cimone, Italy): annual and seasonal variability. AAQR. doi:10.4209/aaqr.2015.05.0364

Lo Vullo E, Furlani F, Arduini J et al (2016b) Anthropogenic non-methane volatile hydrocarbons at Mt. Cimone (2165 m a.S.L., Italy): impact of sources and transport on atmospheric composition. Atmos Environ 140:395–403

Miller BR, Weiss RF, Salameh PK et al (2008) Medusa: a sample preconcentration and GC/MS detector system for in situ measurements of atmospheric trace halocarbons, hydrocarbons, and sulfur compounds. Anal Chem 80:1536–1545

Monks PS, Granier C, Fuzzi S et al (2009) Atmospheric composition change – global and regional air quality. Atmos Environ 43(33):5268–5350

Morgan WT, Allan JD, Bower KN, Highwood EJ, Liu D, McMeeking GR, Northway MJ, Williams PI, Krejci R, Coe H (2010) Airborne measurements of the spatial distribution of aerosol chemical composition across Europe and evolution of the organic fraction. Atmos Chem Phys 10:4065–4083. doi:10.5194/acp-10-4065-2010

Novelli PC (1999) CO in the atmosphere: measurement techniques and related issues, Chemosphere. Global Change Sci 1:115–126

Novelli PC, Masarie KA, Lang PM et al (2003) Reanalysis of tropospheric CO trends: effects of the 1997–1998 wildfires. J Geophys Res. doi:10.1029/2002JD003031

Parrish DD, Allen DT, Bates TS, Estes M, Fehsenfeld FC, Feingold G, Ferrare R, Hardesty RM, Meagher JF, Nielsen-Gammon JW, Pierce RB, Ryerson TB, Seinfeld JH, Williams EJ (2009) Overview of the Second Texas Air Quality Study (TexAQS II) and the Gulf of Mexico Atmospheric Composition and Climate Study (GoMACCS). J Geophys Res 114:D00F13. doi :10.1029/2009JD011842

Parrish DD, Law KS, Staehelin J et al (2012) Long-term changes in lower tropospheric baseline ozone concentrations at northern mid-latitudes. Atmos Chem Phys 12:11485–11504

Petritoli A, Giovanelli G, Kostadinov I (2003) SCIAMACHY validation of NO2 total column by means of ground-based DOAS measurements at Mt. Cimone (44N, 11E) and Stara Zagora (45N, 25E) stations. In: Proceedings of Envisat Validation Workshop, Frascati, Italy, 9–13 December 2002 (ESA SP-531, August 2003)

Petritoli A, Bonasoni P, Giovanelli G et al (2004) First comparison between ground-based and satellite-borne measurements of tropospheric nitrogen dioxide in the Po basin. J Geophys Res. doi:10.1029/2004JD004547

Putero D, Cristofanelli P, Sprenger M et al (2016) STEFLUX, a tool for investigating stratospheric intrusions: application to two WMO/GAW global stations. Atmos Chem Phys. doi:10.5194/acp-16-14203-2016

Ramachandran S (2015) New directions: mineral dust and ozone – heterogeneous chemistry. Atmos Environ 106:369–370

Real E, Law KS, Weinzier B et al (2007) Processes influencing ozone levels in Alaskan forest fire plumes during long-range transport over the North Atlantic. J Geophys Res. doi:10.1029/2006JD007576

Roscoe HK, Johnston PV, Van Roozendael M et al (1999) Slant columns measurements of O3 and NO2 during the NDSC intercomparison of zenith-sky UV-visible spectrometers in June 1996. J Atmo Chem 32:281–314

Saliba M, Ellul R, Camilleri L et al (2008) A 10-year study of background surface ozone concentrations on the island of Gozo in the Central Mediterranean. J Atmo Chem 60:117–135

Schultz MG, Akimoto H, Bottenheim J et al (2015) The global atmosphere watch reactive gases measurement network. Elem Sci Anth. doi:10.12952/journal.elementa.000067

Škerlak B, Sprenger M, Wernli H (2014) A global climatology of stratosphere–troposphere exchange using the ERA-Interim data set from 1979 to 2011. Atmos Chem Phys 14:913–937. doi:10.5194/acp-14-913-2014

Sprenger M, Wernli H, Bourqui M (2007) Stratosphere troposphere exchange and its relation to potential vorticity streamers and cutoffs near the extratropical tropopause. J Atmos Sci 64:1587–1602

Steinbacher M, Zellweger C, Schwarzenbach B et al (2007) itrogen oxide measurements at rural sites in Switzerland: Bias of conventional measurement techniques. J Geophys Res. doi:10.1029/2006JD007971

Stevenson DS, Dentener F, Schultz M et al (2006) Multimodel ensemble simulations of present-day and near-future tropospheric ozone. J Geophys Res. doi:10.1029/2005JD006338

Stohl A, Bonasoni P, Cristofanelli P et al (2003) Stratosphere-troposphere exchange: a review, and what we have learned from STACCATO. J Geophys Res. doi:10.1029/2002JD002490

Stohl A, Forster C, Frank A et al (2005) Technical note: the Lagrangian particle dispersion model FLEXPART version 6.2. Atmos Chem Phys. doi:10.5194/acp-5-2461-2005

Tarasova OA, Brenninkmeijer CAM, Jöckel P et al (2007) A climatology of surface ozone in the extra tropics: cluster analysis of observations and model results. Atmos Chem Phys. doi:10.5194/acp-7-6099-2007

The Royal Society (2008) Ground-level ozone in the 21th century: future trends, impacts and policy implications. The Royal Society, London

UNEP and WMO (2011) Integrated assessment of black carbon and tropospheric ozone. UNEP, Nairobi

Val Martin M, Honrath RE, Owen RC (2006) Significant enhancements of nitrogen oxides, black carbon, and ozone in the North Atlantic lower free troposphere resulting from North American boreal wildfires. J Geophys Res. doi:10.1029/2006JD007530

Wagner A, Blechschmidt A-M, Bouarar I et al (2015) Evaluation of the MACC operational forecast system – potential and challenges of global near-real-time modelling with respect to reactive gases in the troposphere. Atmos Chem Phys. doi:10.5194/acp-15-14005-2015

Westerling AL, Hidalgo HG, Cayan DR et al (2006) Warming and earlier spring increase western US forest wildfire activity. Science 313:940–943

Zanis P, Schuenpbach E, Gaeggeler HW et al (1999) Factors controlling beryllium-7 at Jungfraujoch in Switzerland. Tellus 51(4):789–805

Zanis P, Hadjinicolaou P, Pozzer A et al (2014) Summertime free-tropospheric ozone pool over the eastern Mediterranean/Middle East. Atmos Chem Phys. doi:10.5194/acp-14-115-2014

Zellweger C, Ammann M, Buchmann B et al (2000) Summertime NOy speciation at the Jungfraujoch, 3580 m above sea level, Switzerland. J Geophys Res 105(D5):6655–6667

Zellweger C, Huglin C, Klausen J et al (2009) Inter-comparison of four different carbon monoxide measurement techniques and evaluation of the long-term carbon monoxide time series of Jungfraujoch. Atmos Chem Phys. doi:10.5194/acp-9-3491-2009

Zellweger C, Steinbacher M, Buchman B (2012) System and performance audit of surface ozone, Methane, Carbon Dioxide, Nitrous Oxide and Carbon Monoxide at the Global GAW Station Mt. Cimone, Italy, September 2012

Studies on Environmental Radionuclides at Mt. Cimone

Abstract Naturally-occurring and artificial radionuclides in PM_{10} aerosol samples have been systematically measured for more than 12 years at the "O. Vittori" observatory, a baseline station located in a crucial position in the middle of the Mediterranean basin under the influence of relevant atmospheric streamers crossing in this area. The database collected and herein described covers PM_{10} mass load, ^{7}Be and ^{210}Pb, while sporadic samples showed the occurrence of artificial radionuclides as a result of accidents such as the tsunami induced Fukushima nuclear accident or the Algeciras non-nuclear ^{137}Cs release. The principal scope of radioactivity monitoring at CMN has been the study of Stratosphere to Troposphere Exchange, mainly based on the variation of cosmogenic ^{7}Be, while the availability of ^{210}Pb as well of the mass load of PM_{10} allowed to extend the efficiency of radiotracer data to the identification of continental aerosol sources such as Saharan dust and the Balkan region. Extensive work has been and is still in progress based on the collected dataset concerning the application of global circulation models which rely on the support of objective tracers for their validation.

Keywords ^{7}Be,^{210}Pb • PM_{10} • Stratosphere-to-troposphere exchange • Back-trajectories • Atmospheric modelling • Saharan dust • Fukushima • Algeciras

1 Introduction

The application of radiotracers in the field of atmospheric science dates back to several decades ago, but is still a very powerful tool for a series of reasons.

This approach, mainly born for radioprotection purposes at the time of Cold War, when atmospheric nuclear testing brought to light the effectiveness of atmospheric circulation in redistributing pollutants and in particular radioactivity (still perceived as the most threatening among them) at the global scale, is still very appreciated and used, in spite of its niche application.

Most noticeably, the Cold War outcome from nuclear weapon testing posed the conceptual basis for what is now defined as "biogeochemistry", i.e. the science of the dynamics of matter (both natural and anthropogenic) including both transport and transformation within a mass balance conceptual scheme; moreover in this

P. Cristofanelli et al., *High-Mountain Atmospheric Research*, SpringerBriefs in Meteorology, DOI 10.1007/978-3-319-61127-3_4

framework at least two fundamentals of the modern environmental science have been introduced and developed, namely the design of monitoring network for the observation of environmental data, now widely extended to a series of stable pollutants and the concept of exposure in relation to health risk assessment, a term which now pertains, for example, also to ozone (O_3) and to airborne particulate.

Along with the assessment of massive artificial radioactivity recirculated by nuclear weapon testing, several natural radionuclides, i.e. ^{7}Be and ^{210}Pb for the aerosol phase and ^{222}Rn for the gaseous one, were detected and their use has long been recognized as invaluable in atmospheric studies, providing an extremely reliable source of quantitative information on which supporting the interpretation of the complex dynamics of this environmental compartment. More details on the topic can be found in Tositti et al. (2014), Turekian and Graustein (2003) and references therein.

Though the use of radionuclides is greatly appreciated and included even within the WMO[1] framework (WMO-GAW 2001, 2004), their use is often perceived as old-fashioned; nevertheless their niche use in environmental studies is generally extremely informative, sometimes providing the only way of solving complex topographical or dynamical situation, and efficient as witnessed by recent literature (Itoh and Narazaki 2016; Liu et al. 2016; Fontes et al. 2014) and from what will be reported in detail herein.

The main reason to adopt a radiotracer approach in geophysics is based on the following properties:

- Radionuclides (in particular cosmogenic and natural ones) have sources which are very well defined both in term of magnitude/intensity and location;
- They hardly suffer for ambiguous source terms, i.e. they usually have very well defined single sources;
- They are accurately measurable even at very low concentration level;
- Owing to their physical decay rate they may add kinetical information on geophysical processes;
- When artificial radionuclides emitted accidentally to the atmosphere are co-detected in the same environmental samplings, crossing information with natural ones greatly enhances the interpretative capacity of atmospheric (and environmental) processes and of their time evolution.

As a result, environmentally occurring radiotracers are typically used as proxies of other pollutants as they help to build a solid conceptual framework applicable to other environmental species whose budget is especially complex, provided they have very close environmental behaviour (Baskaran 2012; Reiter 1978).

While each environmental compartment may take advantage of a characteristic array of radiotracers, this chapter will obviously focus on the lower atmosphere, including therefore troposphere and stratosphere, where the most abundant and important naturally occurring radionuclides, apart from the gaseous ^{222}Rn ($t_{1/2} = 3.83$ d), are ^{7}Be ($t_{1/2} = 53.3$ d) and ^{210}Pb ($t_{1/2} = 22.1$ y). Differently from ^{222}Rn, both the

[1] World Meteorological Organization.

latter radionuclides are particle-reactive, therefore they populate submicron aerosol particles; moreover they both typically show a single size distribution peak in the accumulation mode (Gründel and Porstendörfer 2004), where airborne particulate is characterized by the highest atmospheric residence time usually in the week or even longer time range (Hobbs 2000), therefore allowing their application up to the meso- to synoptic scales.

The origin of ^{7}Be and ^{210}Pb is respectively cosmogenic and terrestrial. This allows to locate the source of ^{7}Be in the stratosphere and in the upper troposphere, while ^{210}Pb has an indirect source in the continental surface of planet. The rationale for this is their respective mechanism of production/formation. In fact, ^{7}Be depends in a complex way on the interaction, i.e. spallation reactions, between galactic cosmic rays with atmospheric nitrogen and oxygen producing a well- defined vertical concentration profile of the radionuclide peaking in the mid-stratosphere (Usoskin et al. 2009), while ^{210}Pb depends on the emanation of ^{222}Rn from rocks/soils as a result of the ubiquity of $^{238}U/^{226}Ra$ system, subsequent decay of radon into a series of short-lived radioactive intermediates until ^{210}Pb is produced (Baskaran 2011). More detail and references, including historical and up-to-date papers, can be found in Tositti et al. (2014).

As a result, ^{7}Be and ^{210}Pb are historically employed as tracers of air masses originating respectively in the stratosphere/upper troposphere and in the Planetary Boundary Layer (PBL).

In this chapter, we describe the results from a long-term research activity carried out at CMN including over 12 years of data collection in the period 1998–2011. Though the observational activity ceased several years ago, data elaboration, interpretation and application to circulation modelling and model validation is still ongoing in the framework of a several collaborations promoted by the appealing properties of our dataset, fairly large (consisting of more than 1400 ^{7}Be and ^{210}Pb observations), but above all, characterised by a time resolution of about 2 days enabling a satisfying capture of meteorologically driven variability.[2]

2 Historical and Background

The measurement of airborne radioactivity on CMN samples at Bologna University laboratory (Department of Chemistry) started in the mid 90's when preliminary measurements of ^{7}Be were carried out to identify intrusions of stratospheric air-masses at the measurement site. This activity was devoted to assess the contribution of stratospheric O_3 to the overall tropospheric budget pioneered by Paolo Bonasoni at CNR-FISBAT (National Research Council of Italy-Institute for Physics of the

[2]The most widely adopted time resolution in airborne radionuclide monitoring is on a weekly basis, which typically smooths out fluctuation due to meteorological variability leading to a remarkable loss of information.

High and Low Atmosphere). In this framework, occasional TSP[3] aerosol samples from the Meteorological Office of the Italian Air Force hosting the CMN facility dedicated to meteorological and climatological observations since several decades. The limited but successful radioactivity measurements carried out in close collaboration with CNR-FISBAT and the colleagues from the Italian Air Force, brought to the inception of a systematic activity of aerosol sampling and measurements which has been so far unique, very productive and inspiring, though never directly funded by any official research grant or economical support. All the resources available for our experiment have been due mostly to volunteering and a smart management of other research grants for filters and γ-spectrometry, while we acknowledge the generous support in sampling management by the FISBAT/ISAC team and the provision of a sampler which was donated by another CNR institute to this scope.

One of the most innovative characteristic of the data collected has been the introduction of an aerosol sampling approach based on PM_{10} (i.e. mass of aerosol particles with aerodynamic diameter lower or equal to 10 μm) metrics, which in 1997–1998 had not yet been officially activated within the EU Air Quality networks and which was still limitedly applied even within the European aerosol research framework. Therefore, beside the analysis of airborne radionuclides by high resolution γ-ray spectrometry, all the particulate samples collected with our high volume PM_{10} sampler were weighed under controlled conditions and mass load registered contributing to our database.

It is worth noting that, before our research activity, radioactivity has been measured at CMN for many decades, mostly within the radioprotection monitoring surveillance of atmospheric weapon testing which produced the emission and redistribution of artificial radionuclides at the global scale. Within this framework, the Italian Air Force collected total β radioactivity in small aerosol filters for many decades; a comprehensive database was collected and reorganized in an inclusive report by CNR-IFA (Italian National Research Council, Institute of Physics of the Atmosphere, Dietrich et al. 1997) soon after the dismissing of the monitoring activity. In addition, a small number of publications on the detection and characterization of the radioactive fallout in Italy following nuclear weapon tests have been reported including on single paper devoted to the French experiment in Algeria Gerboise in 1960 (Argiero et al. 1961). This work reports a table of aerosol radioactivity data from several Italian locations including CMN (Table 1). This chapter therefore anticipates some of the most relevant research topics concerning CMN airshed covering radioactivity measurement, but also Saharan dust incursions implying not only their influence on ambient aerosol behaviour and properties, but also suggesting the importance of radiotracers in atmospheric investigations.

[3] TSP stands for Total Suspended Particulate, i.e. ambient aerosol samples collected without any size cut-off.

Table 1 Locations and counts detected at Italian locations in the aftermath of the Gerboise French atomic explosion (Argiero et al. 1961)

Place	Height (m a.s.l.)	Counts (pC m^{-3})
Pian Rosà	3488	15
Monte Cimone	2173	8
M. Terminillo	1875	0.75
M. Scuro	1716	1.8
M. S. Angelo	844	0.6
Resina	610	0.22
Vigna di Valle	270	0.8
Livorno	3	0.15

3 Methodologies

At CMN station, aerosol mass load in the form of PM_{10} was sampled with a Thermo-Environmental PM_{10} high-volume sampler (average flow rate 1.13 m^3 min^{-1}) on rectangular glass fiber filters (Whatman, 20.3 cm × 25.4 cm) manually changed every 2–3 days (average volume collected on each filter equal to 3000–4000 m^3). After sampling, PM_{10} samples were delivered to the University of Bologna, where mass load was determined gravimetrically by an electronic microbalance after 24 h filter conditioning. After weighing, PM_{10} samples were subjected to non-destructive ©-spectrometry for the determination of atmospheric nuclides. To this aim, two HPGe (Hyper-Pure Germanium) detectors were used: one p-type coaxial detector (Ortec/Ametek) with a relative efficiency of 32.5% and FWHM (Full Width at Half Maximum) 1.8 at 1332 keV for higher energy range (1000–2000 keV), and one planar DSG detector with an active surface of 1500 mm^2 and FWHM of 0.73 at 122 keV for lower energies (10–900 keV). Spectra were accumulated for at least 1 day to optimize peak analysis and then processed with a specific software package (GammaVision32, version 6.07, Ortec). Efficiency calibration was determined on both detectors with a blank glass fiber filter traced with accurately weighed traced with accurately weighted aliquots of a standard solution of mixed radionuclides (QCY48, Amersham) supplemented with ^{210}Pb homogeneously dispersed dropwise over the filter surface. The calibration filter was dried under a hood under ambient conditions and then folded into a polystyrene container in the same geometry as the unknown samples. For quantitative analysis of samples, the spectrum of a blank filter in the same geometry was subtracted, while uncertainty on peaks ($k = 1$, 68% level of confidence) was calculated propagating the combined error over the efficiency fit with the counting error. Further details on measurement techniques are available elsewhere (Tositti et al. 2012; Brattich et al. 2017).

4 Use of Radiotracers in Atmospheric Studies

4.1 Stratosphere-to-Troposphere Exchange Studies

The early activity produced by the Environmental Chemistry and Radioactivity Laboratory of Bologna University was mainly devoted to diagnostic purposes, i.e. the identification of stratosphere-to troposphere exchange events by ^{7}Be, one of the most classical and widely employed applications of atmospheric radiotracers.

Pioneering work on this topic dates back to the 60–80's, when several and basically still valuable papers on Stratosphere-to-Troposphere Exchange (STE) associating radioactivity, O_3 and potential vorticity were published (e.g. Danielsen 1968; Reiter 1975; Viezee and Singh 1980). This integrated approach provides a reference framework enlightening the great potential of ^{7}Be in STE diagnosis. The interest in STE relies on the influence of this phenomenology on O_3 tropospheric budget, since the stratosphere is the major atmospheric reservoir of both ^{7}Be and O_3[4] as previously mentioned. In fact, since O_3 plays a major role in the oxidant capacity of the troposphere and is also a powerful greenhouse gas, accurate estimates of its photochemical production as well as of the stratospheric contribution are of great importance. Owing to the non- obvious phenomenology of STE (Stratosphere-Troposphere Exchange) and to its unpredictable spatial evolution, STE events are successfully investigated at high elevation observatories where their occurrence can be more likely detected.

The need to investigate tropospheric O_3 phenomenology across the European territory was at basis of the two projects EC VOTALP – Vertical O_3 Transports in the Alps I and II; within this framework, O_3 behaviour at CMN was investigated in association with ^{7}Be and potential vorticity. Results from this early activity did not allow for a systematic analysis of the radionuclidic dataset owing to the limited consistency of the samplings collected; nevertheless, in this period some of the highest ^{7}Be concentration peaks of our whole dataset were detected leading to the identification of STE events of especially high intensity as reported in Bonasoni et al. (1999, 2000). In particular, values of 14.8 (on 30 May 1996), of 12.7 (31 May 1996) and 23.6 mBq m^{-3} (on July 8 1997) were detected. The associated STE events were analysed in detail in terms of meteorological conditions and ^{7}Be concentration. For ^{7}Be, fixed threshold values (i.e. 8 mBq m^{-3}) was tentatively used in agreement with Reiter et al. (1983). ^{7}Be concentrations were also used in combination with ^{210}Pb according to the parameterization by Graunstein and Turekian (1996) introduced the concept of normalised fraction $f(^{7}Be, ^{210}Pb) = (^{7}Be)/(^{7}Be + n^{210}Pb)$ where n is approximated by the ratio of the standard deviation of ^{210}Pb in the dataset to eliminate the influence of wet scavenging. Moreover, as detailed in Chapter 3, also in-situ O_3 values and air-mass transport as diagnosed by FLEXTRA back-trajectories (see https://www.flexpart.eu/, page visited January 27, 2017) (Stohl et al. 1995) were used to identify the occurrence of STE events at CMN.

[4] Until the 60's it was believed that the main source of tropospheric ozone was the stratosphere.

In a further paper, an overview of ^{7}Be data collected simultaneously for over 3 years at four high altitude stations in Europe including CMN (44°11′N, 10°42′E, 2165 m a.s.l.; Italy), Sonnblick (47°03′N, 12°58′E, 3150 m a.s.l.; Austria), Junfraujoch (46°33′N, 07°59′E, 3580 m a.s.l.; Switzerland) and Zugspitze (47°25′N, 10°59′E, 2962 m a.s.l; Germany) provided a tool for analysing STE characteristics on a larger than usual spatial scale (Gerasopoulos et al. 2001). Each site dataset was analysed with basic statistical tools and compared while meteorological parameters and back-trajectories were employed to analyse tropospheric dynamics on the local and on the European scale to enlighten similarities and differences. ^{7}Be showed a frequency distribution characterized by two relative maxima, suggesting a bi-modal behaviour and the existence of an uppermost threshold above which STE influence is sensitive at all stations as formerly reported by Reiter et al. (1983). At all sites ^{7}Be showed a clear seasonal variability with higher values in the summer and lower in the winter, while STE confirmed as a very episodic behaviour efficiently captured by tropopause height. 3-D back-trajectory analysis showed that the highest ^{7}Be values are likely attributed to STE in association with an upper troposphere origin over the Northern Atlantic region.

The observation of ^{7}Be allowed to plan and carry out an intercomparison activity among the four radioactivity laboratories involved and coordinated through the VOTALP II project (Tositti et al. 2004). This initiative remains unique so far, but extremely valuable as the project aimed at comparing the behaviour of this radionuclide at different stations based on statistically significant criteria. Data homogeneity among γ-spectrometry labs was tested thanks to the rotation through each participant of filters from the Austrian partner. Given the relatively short half-life of ^{7}Be and to the unavailability of a commercial standard solution for this radionuclide, the round-robin test was carried out using the samples from a huge particulate sampler installed in Vienna for radioprotection purposes. The associated large-sized filters allowed for sectioning each sample in a series of four, in subsequent four portions and each aliquot was circulated among all the units involved for the g-spectra acquisition. Eventually all the results were treated statistically for comparability using the Student's t approach; all the laboratories yielded comparable performances in ^{7}Be metrology, contributing to the overall experimental good quality prior to elaboration and interpretation of field data from the four stations involved.

After the VOTALP experience, the collection of PM_{10}, ^{7}Be and ^{210}Pb data was already established and regularly set up on a bi-daily basis (i.e. the samples were collected continuously for about 48 h) in spite of the need for regular assistance for the manual change of the filters on the high-volume sampler. This favourable situation projected our activity into the following EU Project STACCATO (Influence of Stratosphere-Troposphere Exchange in A Changing Climate on Atmospheric Transport and Oxidation Capacity) aimed at providing a first detailed investigation of STE mixing of stratospheric and tropospheric air and their impact on atmospheric chemistry under climate change. The discriminant for stratospheric O_3 requires the availability of suitable and measurable markers allowing for a reliable distinction of this fraction from the O_3 produced in PBL advected to the mountain tops and/or in situ by local photochemistry. The well characterized and distinct properties of stratospheric air masses provide an array of parameters which altogether converge to minimize the ambiguity in their detection and identification. In general, stratospheric air masses bring about high O_3, ^7Be and potential vorticity,[5] and low relative humidity, therefore the time variation of these parameters toward the described trend is in itself a suitable way of tracing a likely STE event. The progressive increase in the dataset consistency allowed to test on one side the efficiency of this multiparametric approach, but also to refine it through the setup of new numerical tools and finally to investigate on the frequency of STE occurrence both on a seasonal and on a interannual basis (Cristofanelli et al. 2006). In particular, exploiting the time resolution of our sampling approach (2 days against the most widely employed weekly basis which smooths out the fluctuations due to the typical meteorological variability) and on account of the different topo-geographical characteristics of CMN in respect of Zugspitze, we investigated the best way of identifying ^7Be concentration peaks, usually taken as the primary identification tool for STE analysis, firstly based on Reiter's threshold (Reiter et al. 1983) to subsequently develop a more accurate approach. In this light, a dynamic threshold based on the comparison of ^7Be peak with the values collected before and after the potential event date, elaborated through a moving average approach, was introduced. The threshold of 8 mBq m^{-3} used by Reiter et al. (1983) is unlikely to apply tout court to CMN at a lower height and latitude than Zugspitze, where also the Alpine topography might affect differently the atmospheric circulation. For this reason, the running threshold t_D^{Be} as a function of both the ^7Be monthly running mean and the ^7Be residual was calculated as:

$$t_D^{Be} = \overline{^7Be_D} + \frac{\sum_{D=1,N} \left| \overline{^7Be_D} - {}^7Be_D \right|}{N}$$

where $\overline{^7Be_D}$ represents the ^7Be monthly running mean calculated for each day of the sampling month, ^7Be_D is the ^7Be concentration recorded at the Dth sampling and N stands for the number of samplings carried out during the month. In this way ^7Be peaks were selected according to the monthly moving average, as proposed in the paper mentioned above. Among the most relevant results of this study on the basis of the multiparametric approach adopted, there is the observation of a STE seasonality different from what usually believed, but in agreement with previous findings of Reiter et al. (1983) for the Zugspitze. In fact, tropopause folding and stratospheric intrusions, both linked to STE phenomenology, are usually reported in the older literature as more frequent in late winter – spring (Feely et al. 1989; Sprenger and Wernli 2003). These aspects probably need to be investigated in more depth as a function of region/latitude and taking into account the potential influence of

[5]The potential vorticity is defined as the specific volume times the scalar product of the absolute vorticity vector and the gradient of potential temperature. It can be described as the absolute circulation of an air parcel that is enclosed between two isentropic surfaces. Within the troposphere, the values of PV are usually low, while PV rapidly increases from the troposphere to the stratosphere due to the significant change in static stability.

climate change upon atmospheric circulation. In addition, the use of integrated tracers starting from the relative ^{7}Be maxima, were used to classify STE events on the basis of their degree of conservativeness. In fact, as observed by Zanis et al. (1999). STE can be split into direct (fast and vertical downward transport) and indirect (slant and more long-lasting downward transport) events, leading to a minor or major loss of stratospheric properties en-route.

Along with this *quasi* climatological approach, a further elaboration of the radioactivity data with focus on ^{7}Be has inspired the introduction of an empirical parameter defined as a Stratospheric Intrusion Index SI^2 for the identification of deep stratospheric intrusions (Cristofanelli et al. 2009). This class of events traceable to the previously defined "direct" STE events, is rather infrequent, though they are strongly irreversible and may reach ground level. To the scope, a function of the experimental data of ^{7}Be, O_3 and relative humidity weighting and normalizing their deviations from their corresponding monthly running means was elaborated allowing for a pragmatic and efficient identification of outstanding STE events. The use of the methodology firstly introduced by (Cristofanelli et al. 2006) using in situ ^{7}Be measurements together with relative humidity, potential vorticity and total column O_3 also enabled to investigate the influence of STE events on the long-term O_3 variability at CMN (Cristofanelli et al. 2015).

4.2 Applications of Radiotracer Techniques to Regional Circulation Understanding

As previously mentioned, our database includes not only ^{7}Be but also ^{210}Pb and their aerosol carrier matrix collected as PM_{10}. As explained in the introduction the conjunct use of these three parameters have proved to be very efficient in producing and reproducing, when modelling tools are used, the fundamental dynamical features of atmospheric circulation at a given location. In fact, while ^{7}Be is a tracer of upper troposphere and lower stratosphere (from now on UT-LS), ^{210}Pb and PM_{10} are associated to a continental, i.e. PBL, origin. Figure 1 depicts the basic atmospheric behaviour of ^{7}Be, ^{210}Pb and O_3 as related to CMN studies.

The simultaneous analysis of these three parameters at CMN, i.e. at more than 2000 m a.s.l., constitutes therefore an important tool to assess the degree of vertical motion and consequent mixing within the troposphere. This means that they act as proxies of other chemical substances whose origin is less straightforward and/or is affected by complex dynamics not only due to transport, but also to non-linear chemistry, such as in the case of O_3, a secondary pollutant characterized by multiple sources and affected by multiple sinks leading to a globally-averaged tropospheric lifetime of about 23 days (Young et al. 2013). Typically, time variations of O_3 at a given high-elevation site may depend on a downward transport from UT-LS traceable by ^{7}Be, while O_3 photochemically produced within PBL can be traced by a

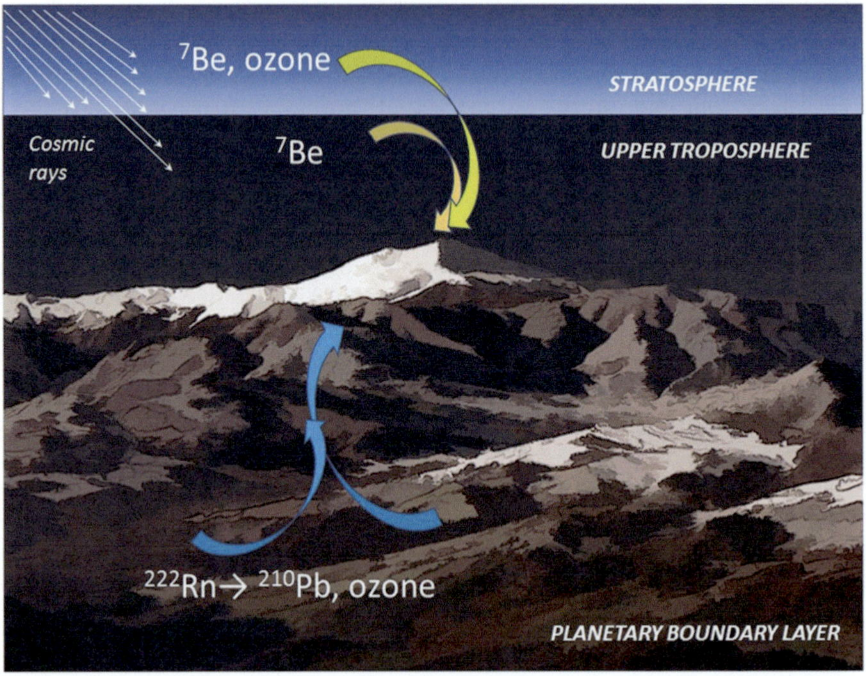

Fig. 1 Atmospheric transport of ^{7}Be, ^{210}Pb and O_3 at CMN station (Image by L. Tositti)

concurrent ^{210}Pb. The availability of simpler but distinct atmospheric species therefore promote data and events interpretation.

In the attempt to assess the continental O_3 sources, an analysis of the radionuclide dataset and of their activity ratio (^{7}Be/^{210}Pb) was carried out by Lee et al. (2007) comparing their behaviour at the two WMO-GAW mountain stations of CMN (Italy) and Mt. Waliguan (36°17′N, 100°54′E, 3816 m a.s.l.; Popular Republic of China). The conjunct use of both radiotracers led to assess that while O_3 at Mt. Waliguan is mainly of stratospheric origin, at CMN it is mainly photochemically produced. Moreover, at both stations O_3 as well as ^{7}Be/^{210}Pb activity ratio show a bi-modal frequency distribution, though characterized by extremely different values, suggesting that in both cases two main processes seem to drive and control O_3 cycling and suggesting the efficacy of the radiotracer approach in understanding the complexity of tropospheric O_3 and in particular in event classification.

The analysis of the whole ^{7}Be, ^{210}Pb and ^{7}Be/^{210}Pb activity ratio collected all over the 12-years period allowed for a much better understanding of processes controlling their seasonality (Tositti et al. 2014). The seasonal cycle of both ^{7}Be and ^{210}Pb radiotracers is characterized by a summer maximum (Fig. 2), even though as for ^{7}Be secondary maxima during the cold/transition months are also present in ^{7}Be seasonality.

The bimodality previously detected in ^{7}Be/^{210}Pb frequency distribution can be associated to the occurrence of a summer maximum (related to the specular sea-

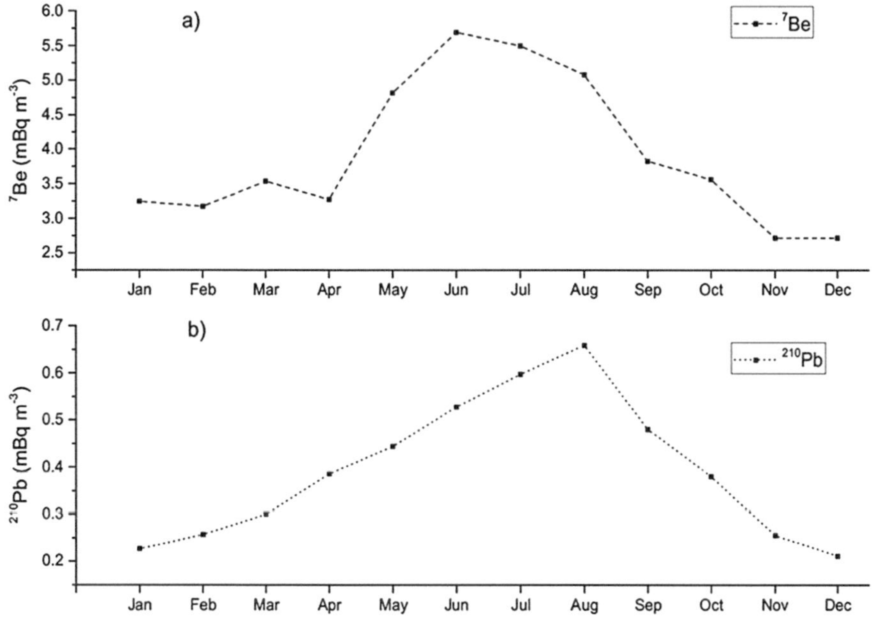

Fig. 2 Seasonal cycle of ^{7}Be (**a**) and ^{210}Pb (**b**) radiotracers at CMN during the sampling period 1998–2011, represented as multi-year monthly means (Adapted from Tositti et al. 2014)

sonal behavior of ^{7}Be and ^{210}Pb) and to the frequent occurrence of peak values during other seasons. On the other hand, ^{210}Pb frequency distribution at CMN strictly follows a log-normal distribution (Fig. 3b). The two ^{7}Be modes (Fig. 3a) may be caused by different physical processes, namely convection during summer and active STE during winter, which were further investigated both through the use of correlation analysis with meteorological parameters as well as by means of a simple mathematical test proposed by Gerasopoulos et al. (2001). Indeed, while during winter a significant inverse correlation of ^{7}Be with relative humidity indicates the role of downward transport from the UT-LS region, in the summer the high correlation among air temperature, ^{210}Pb and ^{7}Be points to the influence of enhanced convective vertical mixing both on ^{210}Pb uplift from the PBL and on the counterbalancing downward ^{7}Be-rich motion from the upper troposphere.

The relative roles of wet scavenging, dry deposition and convective transport and scavenging (coupled) were further investigated with the help of the GMI (Global Modeling Initiative) chemistry and transport (CTM) 3-D model (Brattich et al. 2017), driven by NASA's MERRA (Modern Era Retrospective analysis for Research and Applications). The GMI is a state-of-the-art CTM capable of simulating atmospheric radionuclides ^{222}Rn, ^{210}Pb, ^{7}Be, and ^{10}Be throughout the troposphere and stratosphere (Considine et al. 2004; Liu et al. 2016). The GMI model correctly reproduced ^{210}Pb seasonal cycle at CMN, while ^{7}Be was correctly reproduced only during the spring-winter period, with an important ^{7}Be underestimate during the warm months probably due to the sensitivity to spatial sampling in the model and to

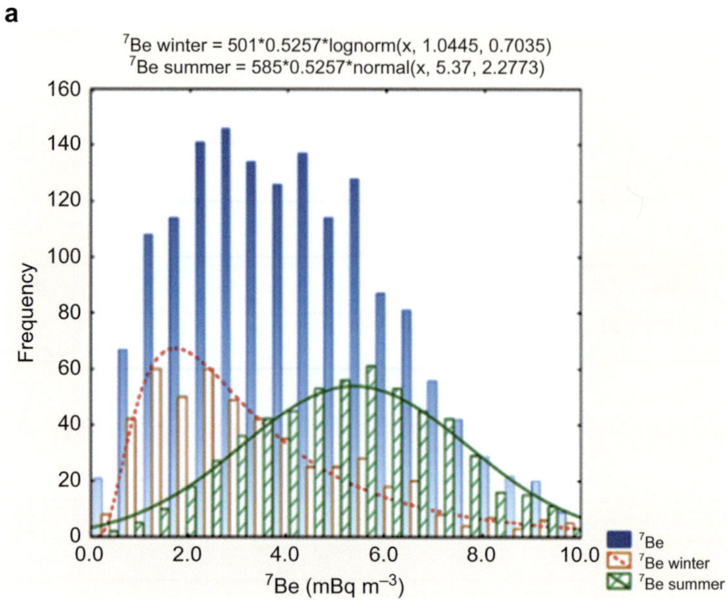

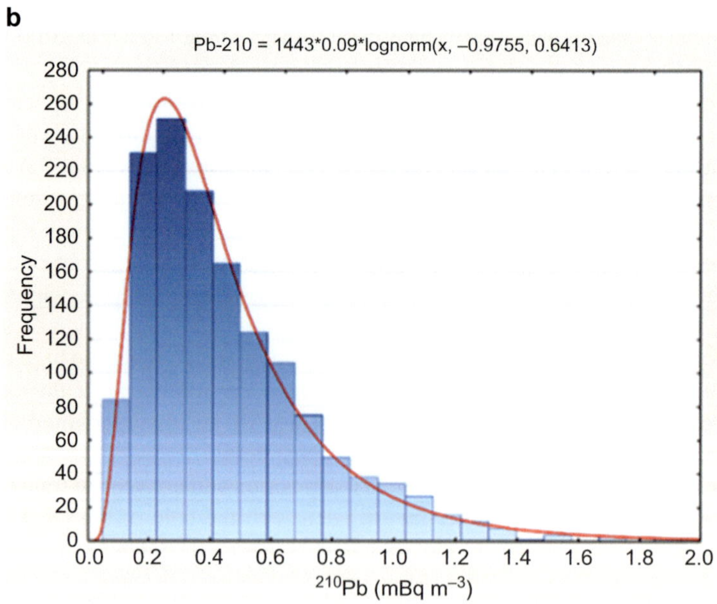

Fig. 3 Frequency distribution of: (**a**) ^{7}Be activities during the winter (November, December, January and February) and summer (May, June, July, and August) seasons at CMN station, fitted respectively by a log-normal and a normal distribution; (**b**) ^{210}Pb activity at CMN station, fitted by a log-normal distribution (Adapted from Tositti et al. 2014)

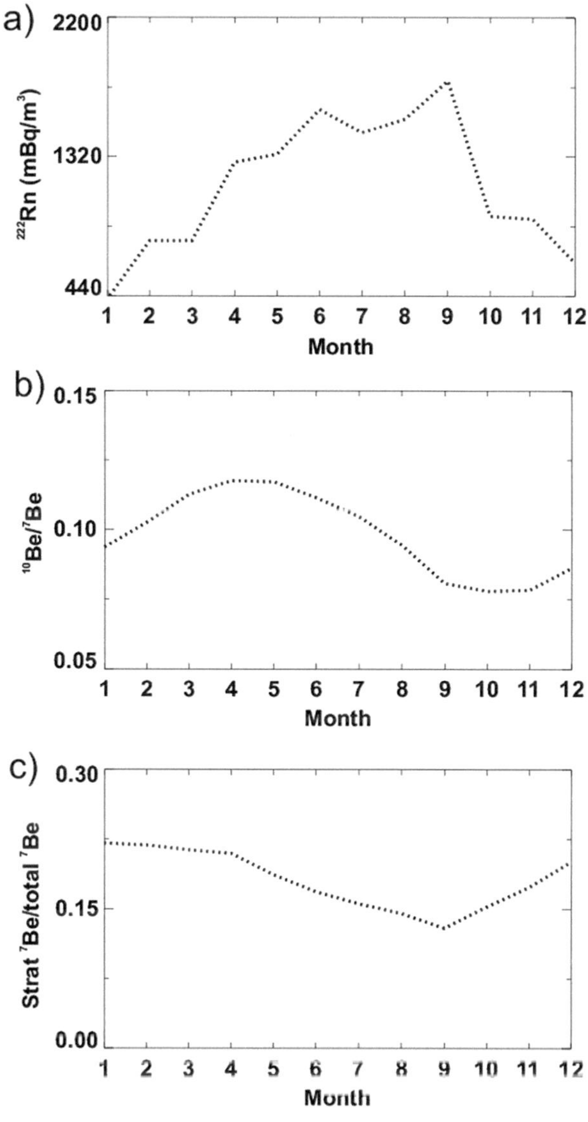

Fig. 4 GMI-simulated seasonal monthly (**a**) ^{222}Rn activities, (**b**) ^{10}Be/^{7}Be ratios, (**c**) stratospheric ^{7}Be/total ^{7}Be ratios (Adapted from Brattich et al. 2017)

the possible incorrect simulation of the mixing between the PBL and the lower free troposphere in this season, when the CMN site can be directly influenced by PBL air masses (Fischer et al. 2003). However, the possibility of simulating seasonal patterns of ^{222}Rn (^{210}Pb parent nuclide) and the fraction of ^{7}Be deriving from the stratosphere (stratospheric ^{7}Be/total ^{7}Be) as well as ^{10}Be/^{7}Be ratio (Fig. 4) enabled to definitively attribute the summer ^{210}Pb maximum with enhanced convective mixing and more ^{222}Rn uplift from the PBL, and the spring-winter ^{7}Be secondary maximum with higher stratospheric influence.

Furthermore, model sensitivity experiments where dry deposition, convection and wet scavenging were separately turned off were used to study and quantify the effect of the different physical processes on ^{7}Be and ^{210}Pb seasonality. The results of sensitivity experiments indicate that scavenging is the most important process controlling the ^{7}Be and ^{210}Pb seasonal pattern, while dry deposition and convective transport/scavenging do not show significant effects.

^{210}Pb peak activities detected along the 12 years' time series were also fully characterized (Brattich et al. 2015a) in terms of meteorological conditions, air mass origin and links with other atmospheric species. As expected, peak ^{210}Pb events mainly occurred during the warm months during periods of prolonged anticyclonic conditions, high temperatures and low relative humidity values, but also strong coupling between the top of the PBL and the free-troposphere, favouring vertical uplift of ^{222}Rn and its progeny from the PBL. ^{210}Pb, together with its daughter nuclide ^{210}Po, is the natural radionuclide with the highest contribution to the annual effective dose received by the population from ingestion and inhalation of uranium and thorium series other than ^{222}Rn (UNSCEAR 2008). For this reason, the average dose increase during the ^{210}Pb peak events at CMN was calculated: interestingly, the average dose increase during selected events represent only a small fraction of the total dose deriving from all sources (natural, medical, ...), even though it might be necessary to evaluate the average dose increase at ground level where ^{210}Pb is emanated together with ^{222}Rn and population density is highest.

4.3 Processes Affecting PM_{10} Variability

In this section we describe the results from the elaboration of our PM_{10} data, which, as previously outlined in the introduction, represent an early but consistent activity collateral to radioactivity observation. Indeed, at the beginning of our experimental activity at CMN, PM_{10} standard, even now rarely used within the field of airborne radioactivity, was still limitedly applied in Italy in the aerosol monitoring and observational activities (it was adopted only later both in the air quality and research fields). Moreover, until the mid-90's most of the experimental activity at CMN was devoted to gases.

The data collected therefore are as unique as the radionuclides time series treated here in this chapter. Though PM_{10} measured at CMN is usually remarkably lower than at ground level station, owing to the elevation and the distance from the direct influence of emission sources, occasional samples reveal extremely high mass loadings, sometimes showing peculiar coloring, at times rather grey to black or others with reddish to orange hues. Apart from the aesthetical considerations, a close association exists between PM_{10}, whose sources are all located at the planet surface, and the terrigenous radionuclide ^{210}Pb, whose precursors are other naturally occurring radionuclides present in all rocks and soils, leading to some extent to a common

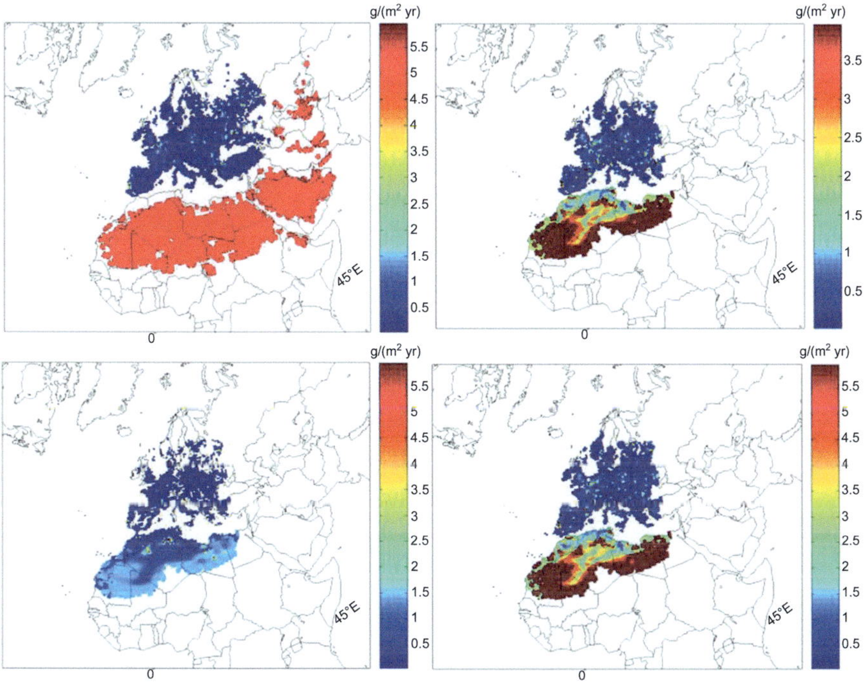

Fig. 5 A-priori (*top-left panel*) and a-posteriori (*top-right panel*) emissions for 2004, estimated by the regularized inverse analysis. In the *bottom* panel the 25% (*left*) and 75% (*right*) percentile probability level. In these figures, emission values lower than 50 mg/(m2 yr) have been omitted (Adapted from Riccio et al. 2009)

environmental fate which led us to investigate their properties and atmospheric behavior.

An early elaboration of PM_{10} dataset was published in 2009 (Riccio et al. 2009) with the aim to understand the main source regions of airborne particulate matter at CMN. The methodology applied was based on an inverse analysis procedure combining the Lagrangian particle dispersion model FLEXPART (Stohl et al. 2005). This analysis allowed to locate the most significant source areas of airborne particulate matter at CMN in the northern African region as well as from northern and north-eastern European countries (Fig. 5). The former contribution was thus associated to Saharan dust outbreaks, which occur the whole year round with a higher frequency in the warm season. In addition, the approach is capable to capture the PM_{10} seasonal behavior characterized by winter minima and summer maxima, enlightening the role of convection in the warm period.

A deeper insight on PM_{10} behaviour at CMN has been provided in (Tositti et al. 2013) who analysed 12 years of data starting from 1998. As previously mentioned and similar to ^{210}Pb, PM_{10} shows a strong seasonal fluctuation, with maxima in the warm season and minima in the cold one as a result of the seasonal pulse of the mixed layer height and of thermal convection leading to strong vertical exchange

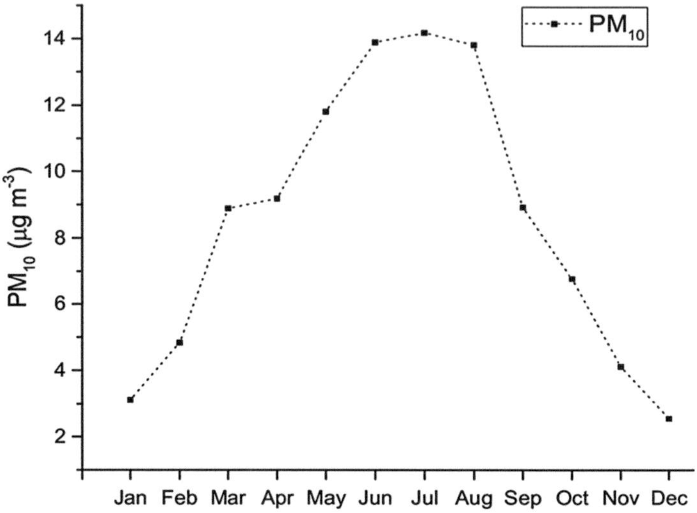

Fig. 6 PM_{10} seasonal cycle at CMN represented as multi-year monthly means

during the warm season (see Fig. 6). The overall mean of PM_{10} over the whole measurement period is 8.8 µg m^{-3}, and its frequency distribution has a log-normal behaviour.

A source-receptor analysis based on back-trajectories clustering technique was used to identify PM_{10} source regions. This allowed to confirm significant contributions from Saharan dust incursions originating in northern Africa, but also important contributions deriving from the uplift of particulate matter produced in the highly polluted Italian regions and from Eastern Europe.

The analysis of PM_{10} data in connection with coarse and fine particle number concentration obtained with an Optical Particle Sizer (OPC Grimm 1.108) operated by CNR-ISAC at the "O. Vittori" station (see Chapter 5), revealed that while during Saharan dust events both the fine and the coarse fraction usually increase significantly and both contribute to PM_{10} mass loading increases, during uplifts of polluted air masses PM_{10} is mainly contributed by submicron particles. In particular, the analysis of the fine vs. coarse particles scatterplot led to the identification of three main clusters of data:

- a large cluster comprehending events characterized by increases in fine particles and small loadings in coarse particles, suggesting event of anthropogenic pollution transport;
- a cluster comprehending events of increases of both the fine and coarse particles, tagged to Saharan dust events;
- a cluster characterized by a huge increase in both fine and coarse particles, almost exclusively belonging to an exceptional Saharan dust event which occurred from 13 to 15 March 2004, leading to the maximum PM_{10} concentration detected at CMN (80 µg m^{-3}) and previously identified and described by (Riccio et al. 2009) and (Beine et al. 2005) and fully analysed by Brattich et al. (2015b).

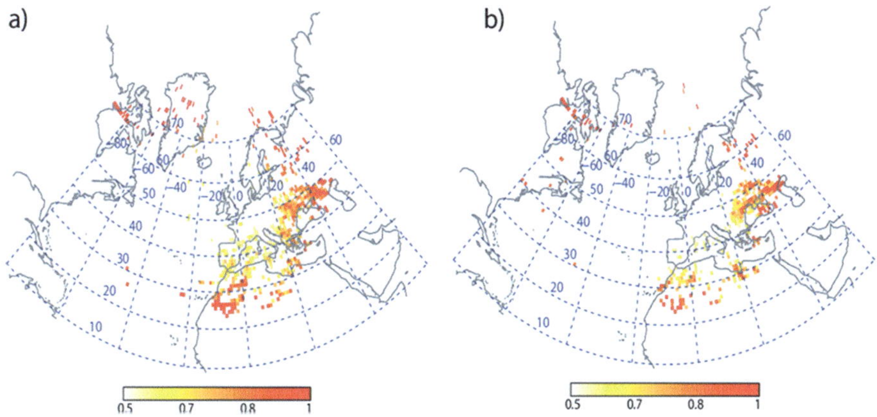

Fig. 7 Source-receptor analysis for high values of PM_{10} (**a**) and ^{210}Pb (**b**) at CMN. Peak events were defined as values higher than 50th percentile (Adapted from Brattich et al. 2016)

The massive transport event of March 2004 was characterized by a peculiar synoptic situation which can be described by a special funnel-shaped feature caused by a steep gradient between an upper level trough extending to low latitudes with minimum centered over North-Western Algeria and a Saharan high extending all over the Mediterranean Sea with an elongated north-eastward tongue (Brattich et al. 2015b). During this event, also increases in ^{210}Pb and ^{7}Be, possibly due to the concurrent dust advection from northern Africa and tropopause folding favoured by the transit of atmospheric trough, were detected.

Since ^{210}Pb is considered not only a tracer of continental air masses (Baskaran 2011), but also a proxy of fine stable aerosol (Pio et al. 2007), we further analysed similarities and differences in ^{210}Pb and PM_{10} at CMN, also by means of trajectory statistical methods. In fact, as shown by Fig. 7 and as highlighted in Tositti et al. (2014), ^{210}Pb and PM_{10} maxima during the warm period are not completely simultaneous, due to the influence from distinct source regions on the two parameters. In fact, as from Fig. 7, Saharan dust advections, because of their strong contributions to coarse particles, affect more sensibly PM_{10} than ^{210}Pb. Other differences arise in the case of European surrounding regions and continental Europe, which again contribute more strongly on PM_{10} than on ^{210}Pb, a behavior suggesting the influence of aged air masses and of particulate of secondary origin (and of biomass burning in the Iberian Peninsula during the summer season), in agreement with (Marinoni et al. 2008). Importantly, through the comparison of results obtained at CMN and at the Huelva coastal low-elevation background site (37°6′N, 6°42′W, 40 m a.s.l.; Spain), this work also shed light on the importance of the Mediterranean as a fine aerosol reservoir layer, again using ^{210}Pb as fine stable aerosol proxy.

4.4 Monitoring of Artificial Radionuclides at CMN

Tropospheric circulation has long been recognised as an extremely efficient mixer of chemicals, regardless of their natural or anthropogenic origin, provided a given chemical species is sufficiently long-lived. In this framework, artificial radionuclides play a leading role not only due to their associated hazard and to the need of early alerting, but also because of their historical role as they opened the road to modern environmental science as described in the introduction. All the times that artificial radioactivity has been detected in the atmosphere the origin is traced back to military activity (weapon test fallout) or to accidents, suggesting the need for a continuous monitoring activity to face potentially fatal emergencies as witnessed by the number of atmosphere-oriented nuclear networks active worldwide for safety/security purposes (see for example the REMRAD network -see http://www.protezionecivile.gov.it/jcms/en/che_cose.wp;jsessionid=D1141C9651FA50909FB855B2342CC570.worker2?request_locale=en&contentId=APP14945, last accessed 3rd February 2017, and CTBTO-preparatory Commission for the Comprehensive Nuclear-Test-Ban Treaty Organization, see https://www.ctbto.org/specials/who-we-are/, last accessed 3 February 2017). In the following, we provide a description of two representative case studies of artificial radionuclide detection at CMN.

4.4.1 Algeciras 1998

In June 1998, high ^{137}Cs values (19.6 and 18.1 μBq m^{-3}) were detected in two samples at CMN (3–4, and 5–8 June 1998, unpublished data). Even though much lower than levels resulting from nuclear accident or nuclear weapon tests, higher than average ^{137}Cs values were detected in several European countries. These relatively high values traced back from an accidental meltdown of a radiotherapy source in a steel mill in Algeciras, near Gibraltar, in southern Spain, on May 30 1998. The accidental release of artificial radioactivity occurred during a fairly weak and persistent summertime pattern, dominated by westerly winds over Gibraltar Strait and southern Spain, as reconstructed by the forward trajectory shown below (Fig. 8) and in agreement with Vogt et al. (1999) and Krysta and Bocquet (2007).

4.4.2 Fukushima

The most recent severe nuclear accident was the Fukushima event in March 2011. As documented in great detail by the International Atomic Energy Agency – IAEA (IAEA 2015), the release of radioactivity to the sea and to the atmosphere was triggered by the tsunami caused by the undersea Great Tohoku earthquake (9.0 magnitude) off the coast of Japan on 11 March 2011. As a result, after the tsunami hit the coast including the Fukushima Daiichi Nuclear Power Plant, the plant suffered for the failure of the cooling system resulting in a level-7 nuclear meltdown and release

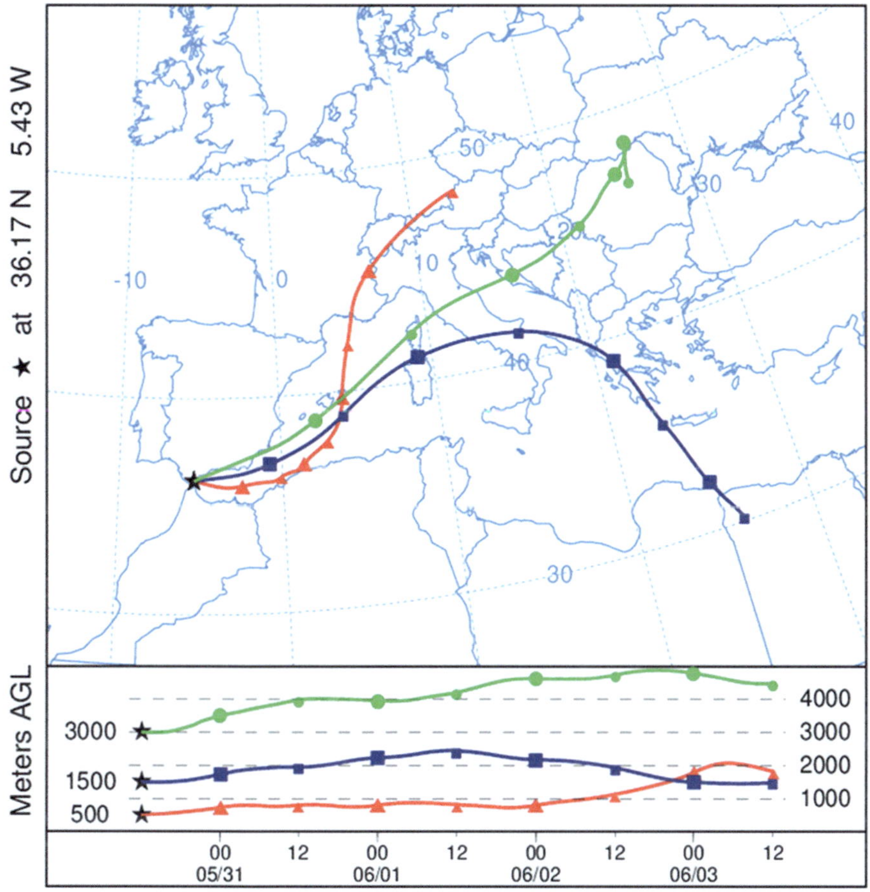

Fig. 8 Forward trajectory starting from Algeciras (36°10′N, 5°26′W) on 30 May 1998 12 UTC, calculated with the HYSPLIT4 model (www.arl.noaa.gov/HYSPLIT.php) at three starting heights (500, 1500 and 3000 m a.g.l.)

of radioactive materials both to the ocean and to the atmosphere. As in the case of Chernobyl accident, the explosion was not originated by the nuclear processes, but by the side production of hydrogen in the hot steel pipes leading to a series of explosions during the following days and consequently to a pulsed multiple emission of fission products to the atmosphere. The global scale alert generated by this huge emergency led this lab to activate a sampling activity both in the Bologna neighborhood and at CMN as reported in Tositti et al. (2012) and Masson et al. (2011). Beyond the naturally occurring radionuclides, the samples collected in the 2011 spring revealed the presence of the artificial radionuclides ^{131}I,^{137}Cs and ^{134}Cs, usually not detected or as in the case of ^{137}Cs, only rarely detected in our samples in the past. Analysis of back-trajectories on the radioactivity data collected confirmed the arrival of artificial radionuclides from Japan coast. Fukushima radioactivity data at

CMN and at Mt. Cuccolino at about 270 m a.s.l. and close to Bologna (60 km north-east from CMN), were mostly comparable, indicating a good vertical mixing of the plume; discrepancies were occasional and related to precipitation events differently affecting the aerosol fate in the troposphere at the two sites. On the basis of the peak ^{137}Cs activity recorded in Fukushima and at CMN, it was estimated that the mean transit time required for the radioactive plume to reach Italy was of the order of 11 days. An interesting outcome of this work was the comparison of the inhalation radiation dose from Fukushima radionuclides and from the natural radionuclides such as ^{7}Be, ^{210}Pb and ^{226}Ra. It was concluded that on an annual basis and on account of the peak Fukushima-related activities, the extra dose estimated in Italy was about 20% of the dose from the natural aerosol radionuclides, which in turn is a about 1% of the overall radiation dose typically absorbed by the Italian population from all the contributing sources. Artificial radioactivity data collected at CMN were also included in a comprehensive paper focusing on Fukushima outcome across the whole European continent (Masson et al. 2011) which, besides being one of the most cited papers on this topic for its exhaustiveness, allowed to show the present high level of awareness and efficiency in the treatment of radioactive plumes, a situation highly improved with respect to the past experience of Chernobyl accident thanks to the remarkable progress in the field of atmospheric sciences. The experience gathered by the Environmental Chemistry and Radioactivity Laboratory of the Bologna University in the field of atmospheric radioactivity and related applications to atmospheric studies also helped a deeper understanding of the transport mechanism of the Fukushima plume to Europe: using data collected at high-altitude European locations, the study of the relation between artificial ^{137}Cs and ^{134}Cs nuclides with ^{7}Be natural nuclide and potential vorticity and relative humidity suggested that maximum artificial radioactivity in Europe was caused by downdraft transport from the UT-LS region (Masson et al. 2016).

Acknowledgments The authors gratefully acknowledge Italian Air Force Meteorological Office (IAFMS) and CNR-ISAC for their precious technical support at the station. CNR-ISAC is gratefully acknowledged for providing infrastructural access at the station. We also acknowledge Dr. Ubaldo Bonafé (CNR-ISAC), Dr. Paolo Bonasoni (CNR-ISAC), Dr. Francescopiero Calzolari (CNR-ISAC), Dr. Paolo Cristofanelli (CNR-ISAC), Dr. Miguel-Angel Hernández-Ceballos (JRC ISPRA), Dr. Giorgia Cinelli, Dr. Hsi-Na Lee (U.S. Department of Homeland Security), Dr. Hongyu Liu (NIA/NASA), Dr. Olivier Masson (IRSN), Prof. Angelo Riccio (University of Napoli Parthenope), and Prof. José Antonio Garcia-Orza (Universidad Miguel Hernandez de Elche) for precious research collaborations. The authors gratefully acknowledge the NOAA Air Resources Laboratory (ARL) for the provision of the HYSPLIT transport and dispersion model used in this publication.

References

Argiero L, Manfredini S, Palmas G (1961) Measurements of air radioactivity in Italy and their relation to the first Sahara atomic explosion. Nature 190:618–619

Baskaran M (2011) Po-210 and Pb-210 as atmospheric tracers and global atmospheric Pb-210 fallout: a review. J Environ Radioact. doi:10.1016/j.jenvrad.2010.10.007

Baskaran M (2012) Handbook of environmental isotope geochemistry. Springer, Berlin Heidelberg

Beine H, Amoroso A, Esposito G et al (2005) Deposition of atmospheric nitrous acid on alkaline snow surfaces. Geophys Res Lett 32:L10808

Bonasoni P, Evangelisti F, Bonafé U et al (1999) Stratosphere-troposphere exchanges: case studies recorded at Mt. Cimone during VOTALP project Phys Chem Earth (C). doi:10.1016/S1464-1917(99)00069-0

Bonasoni P, Evangelisti F, Bonafé U et al (2000) Stratospheric ozone intrusion episodes recorded at Mt. imone during the VOTALP project: case studies. Atmos Environ. doi:10.1016/S1352-2310(99)00280-0

Brattich E, Hernández-Ceballos M, Cinelli G, Tositti L (2015a) Analysis of ^{210}Pb peak values at Mt. Cimone (1998-2011). Atmos Environ 112:136–147

Brattich E, Riccio A, Tositti L et al (2015b) An outstanding Saharan dust event at Mt. Cimone (2165 m a.s.l, Italy) in March 2004. Atmos Environ 113:223–235

Brattich E, Hernández-Ceballos M, JAG O et al (2016) The western Mediterranean basin as an aged aerosols reservoir. Insights from an old-fashioned but efficient radiotracer. Atmos Environ 141:481–493

Brattich E, Liu H, Tositti L et al (2017) Processes controlling the seasonal variations in ^{210}Pb and ^{7}Be at the Mt. Cimone WMO-GAW global station, Italy: a model analysis. Atmos Chem Phys 17:1061–1080

Considine D, Connell P, Logan J (2004) Simulating ozone in the near tropopause region with a new combined model of the stratosphere and troposphere. In: Zerefos C (ed) Quadriennal O3 Symposium, International O3 Commission

Cristofanelli P, Bonasoni P, Tositti L et al (2006) A 6-year analysis of stratospheric intrusions and their influence on O_3 at Mt. Cimone (2165 m above sea level). J Geophys Res. doi:10.1029/2005JD006553

Cristofanelli P, Calzolari F, Bonafè U et al (2009) Stratospheric intrusion index (SI^2) from baseline measurement data. Theor Appl Clim. doi:10.1007/s00704-008-0073-x

Cristofanelli P, Scheel H-E, Steinbacher M et al (2015) Long-term surface ozone variability at Mt. Cimone WMO/GAW global station (2165 m a.s.l, Italy). Atmos Environ 101:23–33

Danielsen E (1968) Stratospheric-tropospheric exchange based on radioactivity, ozone and potential Vorticity. J Atmos Sci 25:502–518

Dietrich E, Favale B, Passamonti V (1997) Trentasei anni di misure di radioattività beta dell'aria sull'Italia al livello del suolo, Valori medi mensili dal 1957 al 1992. IFA, Roma

Feely HW, Larsen RJ, Sanderson CG (1989) Factors that cause seasonal variations in beryllium-7 concentrations in surface air. J Environ Radioactiv. doi:10.1016/0265-931X(89)90046-5

Fischer H, Kormann R, Klüpfel T et al (2003) O_3 production and trace gas correlations during the June 2000 MINATROC intensive measurement campaign at Mt. Cimone. Atmos Chem Phys 3:725–738

Fontes T, Silva M, Silva L et al (2014) Can artificial neural networks be used to predict the origin of O_3 episodes? Sci Tot Environ. doi:10.1016/j.scitotenv.2014.04.077

Gerasopoulos E, Zanis P, Stohl A et al (2001) A climatology of ^{7}Be at four high-altitude stations at the alps and the northern Apennines. Atmos Environ. doi:10.1016/S1352-2310(01)00400-9

Graustein W, Turekian K (1996) ^{7}Be and ^{210}Pb indicate an upper troposphere source for elevated O_3 in the summertime subtropical free troposphere of the eastern North Atlantic. Geophys Rese Lett 23(5):539–542

Gründel M, Porstendörfer J (2004) Differences between the activity size distributions of the different natural radionuclide aerosols in outdoor air. Atmos Environ. doi:10.1016/j.atmosenv.2004.01.043

Hobbs P (2000) Atmospheric aerosols. In: Hobbs P (ed) Introduction to atmospheric chemistry. Cambridge University Press, Cambridge

IAEA (2015) The Fukushima Daiichi accident. IAEA, Vienna

Itoh H, Narazaki Y (2016) Fast descent routes from within or near the stratosphere to the surface at Fukuoka, Japan, studied using ^{7}Be measurements and trajectory calculations. Atmos Chem Phys. doi:10.5194/acp-16-6241-2016

Krysta M, Bocquet M (2007) Source reconstruction of an accidental radionuclide release at European scale. Q J R Meteorol Soc. doi:10.1002/qj.3

Lee H, Tositti L, Zheng X et al (2007) Analyses and comparisons of variations of ^{7}Be, ^{210}Pb, and $^{7}Be/^{210}Pb$ with O_3 observations at two global atmosphere watch stations from high mountains. J Geophys Res. doi:10.1029/2006JD007421

Liu H, Considine D, Horowitz L et al (2016) Using beryllium-7 to assess cross-tropopause transport in global models. Atmos Chem Phys. doi:10.5194/acp-16-4641-2016

Marinoni A, Cristofanelli P, Calzolar F et al (2008) Continuos measurements of aerosol physical parameters at the Mt. Cimone GAW station (2165 m asl, Italy). Sci Tot Environ 391:241–251

Masson O, Baeza A, Bieringer J et al (2011) Tracking of airborne radionuclides from the damaged Fukushima Dai-Ichi nuclear reactors by European networks. Environ Sci Technol. doi:10.1021/es2017158

Masson O, Bieringer J, Brattich E et al (2016) Variation in airborne ^{134}Cs, ^{137}Cs, particulate ^{131}I and ^{7}Be maximum activities at high-altitude European locations after the arrival of Fukushima-labeled air masses. J Environ Radioac 162-163:14–22

Pio C, Legrand M, Oliveira T et al (2007) Climatology of aerosol composition (organic versus inorganic) at nonurban sites on a west-east transect across Europe. J Geophys Res 112:D23S02

Reiter E (1975) Stratospheric-tropospheric exchange processes. Rev Geophys Space Phys 13(4):459–474

Reiter E (1978) Atmospheric transport processes. Part 4: radioactive tracers. In: U.S. Atomic Energy Commission, Division of Technical Information (ed). Springfield, USA

Reiter R, Munzert K, Kanter H et al (1983) Cosmogenic radionuclides and O_3 at a mountain station at 3.0 km a.s.l. AMGBB 23B:131–160

Riccio A, Chianese E, Tositti L et al (2009) Modeling the transport of Saharan dust toward the Mediterranean region: an important issue for its ecological implications. Ecol Quest. doi:10.2478/v10090–009–0019–7

Sprenger M, Wernli H (2003) A northern hemispheric climatology of cross-tropopause exchange for the ERA15 time period (1979–1993). J Geophys Res. doi:10.1029/2002JD002636

Stohl A, Wotawa G, Seibert P et al (1995) Interpolation errors in wind fields as a function of spatial and temporal resolution and their impact on different types of kinematic trajectories. J Appl Meteorol 34:2149–2165

Stohl A, Forster C, Frank A et al (2005) Technical note: the Lagrangian particle dispersion model FLEXPART version 6.2. Atmos Chem Phys 5:2461–2474

Tositti L, Hübener S, Kanter H et al (2004) Intercomparison of sampling and measurement of ^{7}Be in air at four high-altitude locations in Europe. Appl Radiat Isot. doi:10.1016/j.apradiso.2004.04.003

Tositti L, Brattich E, Cinelli et al (2012) Comparison of radioactivity data measured in PM10 aerosol samples at two elevated stations in northern Italy during the Fukushima event. J Environ Radioac. doi:10.1016/j.jenvrad.2012.01.016

Tositti L, Riccio A, Sandrini S et al (2013) Short-term climatology of PM10 at a high altitude background station in southern Europe. Atmos Env. doi:10.1016/j.atmosenv.2012.10.051

Tositti L, Brattich E, Cinelli G et al (2014) 12 years of 7Be and 210Pb in Mt. Cimone, and their correlation with meteorological parameters. Atmos Env. doi:10.1016/j.atmosenv.2014.01.014

Turekian K, Graustein W (2003) Natural radionuclides in the atmosphere. In: Turekian ECH (ed) Treatise on geochemistry, 2nd edn. Elsevier, Amsterdam

UNSCEAR (2008) Sources and effects of ionizing radiation. Report to general assembly with scientific annexes, volume I, annex B. New York: United Nations

Usoskin ICV, Schmidt F, Leppanen G et al (2009) Short-term production and synoptic influences on atmospheric ^{7}Be concentrations. J Geophys Res. doi:10.1029/2008JD011333

Viezee W, Singh HB (1980) The distribution of beryllium-7 in the troposphere: implications on Startospheric/tropospheric air mass exchange. Geophys Res Lett 7(10):805–808

Vogt P, Pobanz B, Aluzzi F et al (1999) ARAC Modeling of the Algeciras, Spain steel mill Cs-137 release. Technical report UCRL-JC-131330. Lawrence Livermore National Laboratory, Livermore, California

WMO-GAW (2001) Global atmosphere watch measurements guide. https://www.wmo.int/pages/prog/gcos/documents/gruanmanuals/GAW/gaw143.pdf Accessed 23 Jan 2017

WMO-GAW (2004) Report no. 155. http://library.wmo.int/pmb_ged/wmo-td_1201.pdf Accessed 17 Jan 2017

Young P, Archibald A, Bowman K et al (2013) Pre-industrial to end 21st century projections of tropospheric ozone from the atmospheric chemistry and climate model Intercomparison project (ACCMIP). Atmos Chem Phys 13:2063–2090

Zanis P, Schuepbach E, Gaggeler H et al (1999) Factors controlling beryllium-7 at Jungfraujoch in Switzerland. Tellus. doi:10.1034/j.1600-0889.1999.t01-3-00004.x

Aerosol Chemical Composition at the Mt. Cimone WMO/GAW Global Station

Abstract The aerosol chemical composition at Mt. Cimone has been investigated since 2000 during a series of field campaigns. Starting from 2009, systematic measurements on PM_1 filters was conducted as a contribution to the EUSAAR/ACTRIS network. Moreover, first online chemical measurements were conducted at Mt. Cimone (CMN) in summer 2012 using a high-resolution aerosol mass spectrometer (AMS). The analyses show that the concentrations of mineral elements, associated to coarse particles, are largely influenced by long-range transport of desert dust from Africa. The submicron aerosol composition is dominated by sulfate, carbonaceous material and, to a lesser extent, by nitrate. The concentrations of all PM_1 major components followed a yearly cycle with maxima in the summer when the station prevalently resides in the planetary boundary layer (PBL). The organic-to-sulfate ratio is smaller in summer then in winter. A meta-analysis of the whole dataset shows evidence of a decrease of sulfate concentrations in PM_1 from 2000 to 2010. The sources of organic matter at CMN are still subject of investigation, although biogenic secondary organic aerosol seems to play a role in summertime while biomass burning becomes prevalent in winter.

Keywords Aerosol chemical composition • Size-distribution of aerosol chemical compounds • Sulphate aerosol • Water-soluble organic carbon (WSOC) • Biomass burning • NMR spectroscopy

1 Introduction

Systematic aerosol chemical observations at CMN date back year 2000 when the size-segregated aerosol composition was determined for particles of diameter between 50 and 10 μm during the MINATROC campaign (Putaud et al. 2004). The experiment focused on the transport of mineral dust particles from Africa, but the results showed a great diversity of aerosol chemical compositions associated to air masses originating from different geographical regions in Europe. The experiment provided therefore the opportunity to characterize the composition of aerosol particles belonging to the regional background in southern Europe and, by exploiting the information carried by chemical composition, to identify source-contributions

P. Cristofanelli et al., *High-Mountain Atmospheric Research*, SpringerBriefs in Meteorology, DOI 10.1007/978-3-319-61127-3_5

to aerosol mass concentrations. Aerosol mass source apportionment is therefore the main objective of chemical speciation monitoring at background sites. However, there exist trade-offs between the degree of sophistication of the chemical measurements and their applicability at remote observatories in a cost-effective manner. The main caveat of chemical determinations with respect to aerosol physical measurements is that, until a few years ago, there were no technological solutions for automatic, unattended monitoring. Before aerosol chemical speciation monitors (ACSM) were invented (Ng et al. 2011), monitoring was conducted by periodic sampling of aerosols on dedicated substrates followed by laboratory analysis. The dataset discussed in this book originates mainly from filter or impactor samples, but new results obtained by means of modern aerosol mass spectrometer during a recent field campaign at CMN are also presented. Chemical composition data obtained on filter samples still represent the large part of the data sets from high-altitude observatories from all over the world. Such datasets, which often extends far in the past, are of particular importance when searching for long-term atmospheric composition changes (Husain et al. 2004; Cozic et al. 2008). As collection on filters generally require several hours of continuous sampling, time resolution is generally much reduced with respect to aerosol physical measurements. As a consequence, atmospheric processes occurring at short time scales cannot be investigated using these sampling methods, and most analyses have focused on the effects of synoptic-scale variability, like air mass transport at large spatial scales (Bourcier et al. 2012), or on seasonal cycles (Kasper and Puxbaum 1998; Henning et al. 2003). The chemical determinations performed in the past 15 years at CMN aerosol samples make no exception. The originality of this dataset comes from the geographical location of the site, located south of other important high-altitude observatories in the great Alpine region (Hohenpeissenberg, Jungfraujoch, Sonnblick), therefore closer to aerosol sources in the Mediterranean region and North Africa. The other peculiarity of CMN is certainly the proximity to a major European pollution hotspot, the Po Valley.

2 A Compilation of Past PM_1 and PM_{10} Chemical Compositions at CMN

Tables 1a and 1b provide a compilation of aerosol chemical measurements performed from 2000 to 2011 at CMN, i.e, until the most recent set of data subject of publication (Carbone et al. 2014). Observations performed in 2012 using an aerosol mass spectrometer are discussed in a following section. The data summarized in the tables originate exclusively from chemical analysis of filter or impactor samples. Common sampling systems included PM_{10} HiVol samplers (60 m^3/h) or low-vol PM_1 or PM_{10} samplers (2.3 m^3/h) (Marenco et al. 2006). During the MINATROC and AEROCLOUDS field campaigns, multi-stage low-pressure impactors have been also employed (Putaud et al. 2004; Carbone et al. 2010). The impactors were

Table 1a Summary of the concentrations of main inorganic ionic species from field campaign observations carried out at CMN between 2000 and 2010

		NO_3^-	$nssSO_4^=$	NH_4^+
2000 SU[1]	PM_1	0.15 (0.02–2.50)	2.15 (0.76–4.64)	1.00 (0.32–1.92)
	PM_{10}	0.86 (0.37–3.51)	2.39 (0.96–5.30)	1.11 (0.38–2.06)
2002–2004 SP/SU[2]	PM_{10}	0.26 (0.11–0.48)	2.77 (1.52–4.73)	1.33 (0.47–2.76)
2002–2004 AU WI[2]	PM_{10}	0.36 (0.03–4.36)	1.21 (0.12–4.36)	0.32 (0.02–2.86)
2004 SU[3]	PM_1	0.13 (0.09–0.64)	2.10 (1.00–4.00)	0.88 (0.48–1.57)
	PM_{10}	0.70 (0.28–1.31)	3.50 (1.10–6.20)	1.24 (0.44–2.56)
2007 SU[4]	PM_1	0.08 (0.04–1.31)	1.25 (0.61–2.22)	0.61 (0.27–1.08)
	PM_{10}	0.66 (0.07–2.27)	1.44 (0.65–2.83)	0.68 (0.27–1.16)
2008 SU[2]	PM_1	0.30 (0.10–0.78)	1.09 (0.39–1.83)	0.42 (0.16–0.75)
	PM_{10}	0.74 (0.33–1.29)	1.32 (0.78–2.69)	0.43 (0.17–0.76)
2009–2011 WI[5]	PM_1	0.04 (0.03–0.46)	0.19 (0.16–0.49)	0.01 (<0.01–0.30)
2009–2011 SP[5]	PM_1	0.13 (0.02–1.44)	0.44 (0.19–1.98)	0.10 (<0.01–0.35)
2009–2011 SU[5]	PM_1	0.34 (0.12–1.38)	1.13 (0.56–2.90)	0.26 (0.08–0.54)
2009–2011 FA[5]	PM_1	0.10 (0.03–0.43)	0.55 (0.15–1.84)	0.10 (<0.01–0.39)
2010 SP[6]	PM_1	0.32 (0.12–0.85)	1.29 (0.83–2.69)	0.25 (0.14–0.53)
	PM_{10}	0.53 (0.19–1.05)	1.52 (0.88–3.00)	0.25 (0.14–0.53)

Results from the EUSAAR/ACTRIS monitoring program between 2009 and 2011 are also presented. Median and (in parenthesis) 10th and 90th percentiles are reported for the concentrations (in $\mu g/m^3$) of each species

References: 1 (Putaud et al. 2004); 2 (unpublished); 3 (Marenco et al. 2006); 4 (Carbone et al. 2010; Cristofanelli et al. 2009); 5 (Carbone et al. 2014); 6 (Sandrini et al. 2014)

"$nssSO_4^=$" stands for non-sea salt sulfate. *WI* winter, *SP* spring, *SU* summer, *AU* autumn

5-stage Berner type with a PM_{10} inlet and equipped with one Tedlar foil and one aluminum foil on each plate (Matta et al. 2003). The stage cutoffs were 0.05, 0.14, 0.42, 1.2 and 3.5 μm, hence providing size-distributions of the aerosol chemical components from the quasi-ultrafine fraction (0.05–0.14 μm) to the coarse fraction (1.2–3.5 μm and 3.5–10 μm). The samplers were typically operated outdoor in the periods of the year when the terrace of the main building was free of snow and ice. More recently (starting from 2009), a dedicated PM_{10} sampling inlet was built across the roof of the station, connected to filter packs and pumps indoor, therefore allowing aerosol collection throughout the year. The new sampling system is a dichotomous sampler (Model 310 Universal Air Sampler TM) for simultaneous PM_1 and PM1–10 aerosol fractions collection. Flow rate is 18 m^3/h and the samples are collected on 9 cm size, pre-baked quartz-fiber filters. The current database (Tables 1a, b) originates from the results of multiple field campaigns and observation programs. Distinct sampling protocols were adopted during the individual experiments. The sampling schedule changed from one filter per day to two samples per day capturing approximately daytime and night-time hours. Sampling duration was varied across seasons (Carbone et al. 2010). Finally, a program of nocturnal samples collection (19:00 to 9:00 local time, LT) was set on a weekly basis (Carbone

Table 1b Same as for Table 1a but for the carbonaceous species

		TC	WSOC	OC	EC
2000 SU[1]	PM_1	1.60 (0.66–2.75)	1.10 (0.48–1.95)	1.16 (0.51–1.99)	0.47 (0.15–0.78)
	PM_{10}	2.56 (1.09–3.87)	1.57 (0.66–2.52)	1.87 (0.83–3.00)	0.58 (0.21–0.93)
2002–2004 SP/SU[2]	PM_{10}	2.61 (1.66–4.81)	1.54 (0.80–3.02)		
2002–2004 AU WI[2]	PM_{10}	0.84 (0.44–4.07)	0.46 (0.20–2.39)		
2004 SU[3]	PM_1			1.34 (1.07–1.97)	0.16 (0.102–0.25)
	PM_{10}			–	–
2007 SU[4]	PM_1	1.67 (0.70–2.66)	0.62 (0.19–1.30)		
	PM_{10}	2.45 (0.76–3.47)	0.83 (0.19–1.65)		
2008 SU[2]	PM_1	2.35 (1.52–2.93)	1.09 (0.71–1.58)		
	PM_{10}	3.16 (2.30–3.86)	1.29 (0.72–1.97)		
2009-2011WI[5]	PM_1	0.62 (0.44–1.00)	0.46 (0.15–0.74)		
2009–2011 SP[5]	PM_1	0.61 (0.24–1.30)	0.38 (0.09–0.95)		
2009–2011 SU[5]	PM_1	1.60 (0.38–2.46)	1.19 (0.33–1.82)		
2009–2011 AU[5]	PM_1	0.73 (0.44–1.34)	0.51 (0.27–0.97)		
2010 SP[6]	PM_1	1.23 (0.66–1.86)	0.44 (<0.01–0.92)	1.14 (0.78–1.58)	0.09 (0.03–0.33)
	PM_{10}	1.23 (0.66–2.12)	0.52 (0.37–0.92)	2.20 (1.39–3.30)	0.09 (0.03–0.33)

TC total carbon, *WSOC* water-soluble organic carbon, *OC* organic carbon, *EC* elemental carbon. All concentrations in μg of carbon per cubic meter

et al. 2014) as a contribution to the atmospheric monitoring activities organized in the frame of the EUSAAR and ACTRIS EU projects.

The resulting timeline of samples from 2000 to 2011 is rather discontinuous (Tables 1a and 1b), with a relative good measurement coverage of the summer period. The main features in the chemical composition of PM_1 and PM_{10} emerging from the existing database can be summarized as follows:

1. Carbonaceous material and $nssSO_4$ are the main chemical constituents of the aerosol at CMN. It should be noted that also mineral dust is deemed to be responsible for periodic very strong enhancements of PM_{10} concentrations, but such events are only rarely captured by the adopted sampling program for chemical analysis. The concentrations of organic carbon (OC) and elemental carbon (EC) determined at CMN in summer 2004 are of the same order of magnitude with the values reported by Pio et al. (2007), for Sonnblick (3106 m a.s.l.) for summer-

time, with an OC/EC ratio of about 8. The higher EC concentrations reported for summer 2000 at CMN are probably affected by charring of organic matter on the aluminum substrates of the Berner impactors (Putaud et al. 2004), therefore the decreasing trend in the concentrations of EC from 2000 to 2010 in Table 1b is probably only apparent. About 50% of total carbon (TC) is water-soluble (WSOC). Water-soluble organic compounds encompass biomass burning products and especially secondary organic aerosol (SOA) compounds. It should be noted that the measured concentrations of WSOC, OC and TC on quartz-fiber filters can be affected by positive artifacts (adsorption of organic vapors on the filter matrix). Thanks to the high face velocity of the dichotomous sampler (95 cm/s), the fraction of TC accounted for by positive artifacts must be in the lower end (ca. 15%) of the range indicated by Cavalli et al. (2016) for European rural stations. The first measurements performed at the site using an aerosol mass spectrometer (AMS), which are unaffected by positive (adsorption) artifacts, confirmed the dominance of organic compounds in PM_1 aerosols (Rinaldi et al. 2015).

2. The concentrations of all carbonaceous and water-soluble inorganic species exhibit a clear seasonal pattern with maximum in the summer, when the station is directly subjected to the influence of planetary boundary layer (PBL) air which carries particles and reactive gases from lower altitudes, including the Po Valley basin. The concentrations of inorganic ions show a clear minimum in wintertime, in agreement with the results obtained at other high-altitude alpine observatories (Kasper and Puxbaum 1998; Henning et al. 2003), and in partial agreement with the measurements at the Massif Central, where submicron nitrate showed a peak in spring months (Bourcier et al. 2012). At CMN, the concentrations of carbonaceous species vary to a much lesser extent between fall, winter and spring with respect to inorganic ions, and, as a consequence, wintertime aerosols exhibit the largest enrichment of carbonaceous species (Carbone et al. 2014), in contrast with the observations from Puy-de-Dôme and Sonnblick showing no organic enrichment in winter (Pio et al. 2007).
3. The comparison of the summertime chemical composition between 2000 (Putaud et al. 2004), 2002–2004 (HiVol samples, unpublished), 2004 (Marenco et al. 2006), 2007 (Carbone et al. 2010), 2008 (unpublished) and 2009–2011 (Carbone et al. 2014) show no evident variation in the atmospheric loadings of the carbonaceous species (Table 1b). By contrast, the non-seasalt-sulfate ($nssSO_4$) concentrations decreased from the MINATROC campaign to the ACTRIS monitoring phase (Table 1a; Fig. 1). Such decrease is particularly evident in the subset of samples collected in daytime, when the impact of PBL air is stronger, indicating that historical changes in emission type/intensity in nearby source regions, and especially in the Po Valley, could have played a major role in the observed negative trend in sulfate concentrations. At the high altitude station of Jungfraujoch, in the Swiss Alps, the analysis of the 1999–2006 record of chemical analysis in TSP does not show any decline in sulfate concentrations (Cozic et al. 2008). Therefore, the decrease observed at CMN must be related to changes in SO_2 emissions in geographical areas south of the Alps.

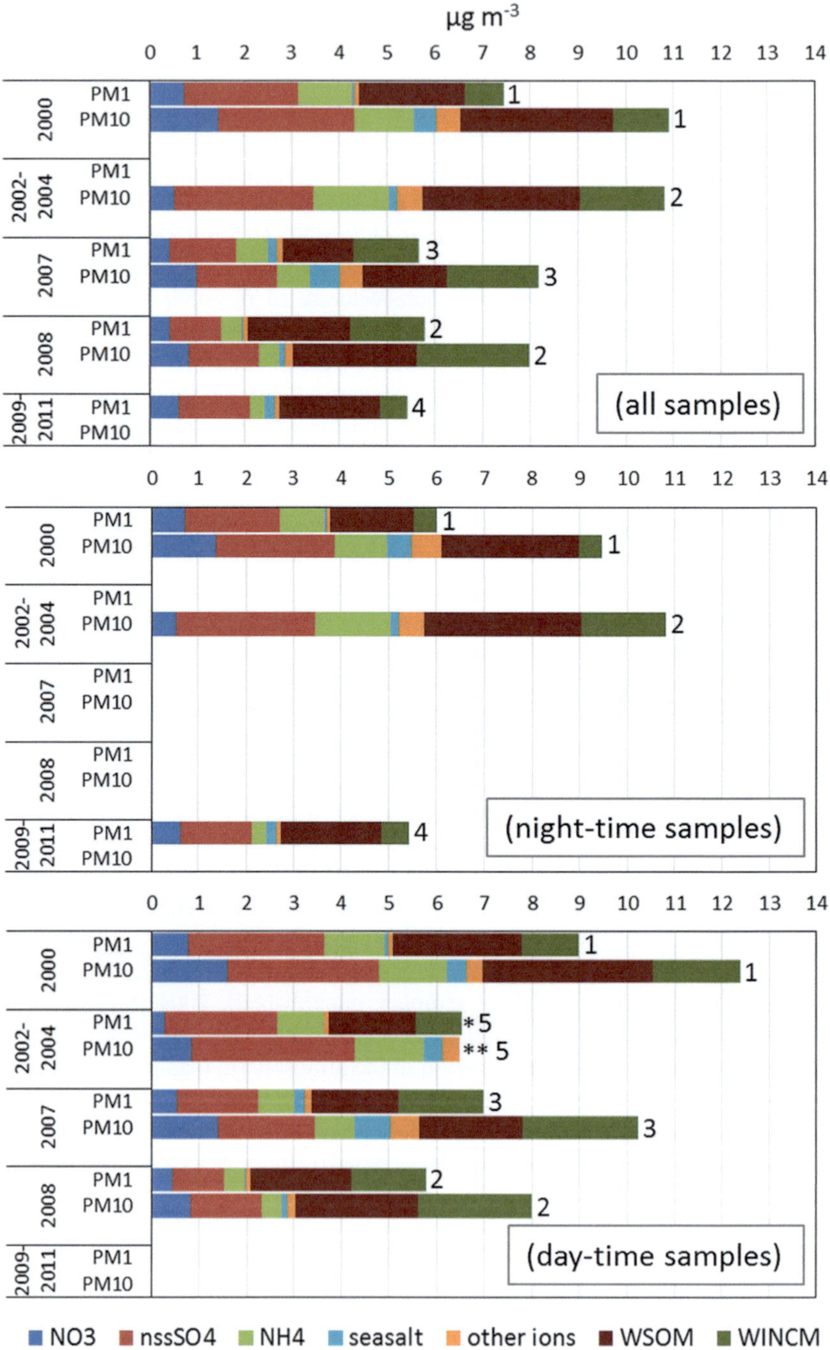

Fig. 1 Mean chemical composition of PM_1 and PM_{10} at CMN during the summer season (JJA). The three panels report data for all sample types (*top*), or specifically night-time samples (*middle*), and daytime samples (*bottom*). Daily samples (24 h) were also considered among the daytime samples (*bottom*) because affected by valley breezes. The data originate from evolved gas analysis

Carbone et al. (2014) demonstrated that the PM_1 concentrations at CMN under conditions of low wind speed in summertime are very well correlated with specific humidity, a tracer of PBL air. The concentrations of submicron nitrate tend to increase during stagnation periods, indicating that their precursors (nitric acid and N_2O_5) originate mainly from PBL sources located in the proximity of CMN (like in the Po basin). However, outside stagnation periods and in other seasons, the PM_1 concentrations at the site are primarily controlled by long-range transport and air mass history. Back-trajectory analysis for the sampling days of ACTRIS monitoring showed that in the warm season, high concentrations of submicron aerosol compounds are found not only during stagnation periods but also for air masses originating from east Europe (Carbone et al. 2014). Such air masses are particularly enriched in submicron sulfate. Interestingly, Putaud et al. (2004), who measured at the site 10 years earlier, did not report any enhancement in PM_1 concentrations for eastern European air masses, which might be explained by the more depressed economies of Eastern European countries at that time. Another air mass type, identified by both elaborations of the MINATROC and ACTRIS datasets, originates from the marine boundary layer of the western Mediterranean basin and is characterizes submicron aerosols with a low nitrate-to-sulfate ratio. The small content of nitrate with respect to sulfate of submicron particles from the Mediterranean basin is confirmed by the observations at the Puy-de-Dôme (Bourcier et al. 2012).

3 Size-Segregated Aerosol Chemical Composition

First size-resolved chemical compositions of PM were determined during the summer 2000 MINATROC campaign (Putaud et al. 2004). However, the peculiar characteristics of the size-segregated aerosol chemistry at CMN were more clearly identified during the following AEROCLOUDS campaigns, when multistage impactor measurements were conducted simultaneously at CMN and at two Po Valley stations: in Milan, and at a rural background site in the southern sector of the Valley (ca. 100 km NE from CMN), San Pietro Capofiume (SPC). Impactor samples for off-line ion chromatography and evolved gas analysis were collected during two distinct short field campaigns, in summer and in winter, and separately between day and night conditions (Carbone et al. 2010). In winter, the concentrations at CMN are one order of magnitude smaller than at the low altitude sites in the Po Valley because

Fig. 1 (continued) and ion chromatography. The seasalt fraction was estimated using the approach of Marenco et al. (2006). "other ions" include potassium and calcium. Water-soluble organic matter (WSOM) was calculated as WSOC*1.9, while water-insoluble carbonaceous matter (WINCM) was estimated as WINC*1.4 (References: 1 (Putaud et al. 2004); 2 (unpublished); 3 (Carbone et al. 2010; Cristofanelli et al. 2009); 4 (Carbone et al. 2014); 5 (Marenco et al. 2006). * Based on measured OC + EC data and WSOC/TC ratios from PM_{10} HiVol data for 2002–2004. ** Carbon data non available) atmospheric stratification traps pollution aerosols in the lower levels. In spite of that, the size distribution of the main chemical species is very similar between CMN and SPC, with a large fraction of the mass of carbonaceous and inorganic ionic compounds associated to accumulation mode particles and much smaller contributions from the smallest (PM0.14) and biggest (PM3.5–10) size ranges. The main difference between CMN and SPC stands in the greater mass fractions of water-insoluble carbonaceous material (WINCM) at the Po Valley sites. In summertime, the concentrations of total reconstructed PM at CMN is still six times lower than in SPC during night-time, while concentrations in daytime are approximately the same at the two stations. Strong convective mixing in daytime hours in the summertime tends to homogenize the aerosol composition in the lower troposphere. This is particularly evident for accumulation mode particles of diameter between 0.14 and 1.2 μm, while the concentrations of coarse aerosols and of quasi-ultrafine particles are lower at CMN than in SPC (Carbone et al. 2010). In daytime, submicron nitrate exhibits higher concentrations at CMN with respect to SPC, which can be explained by the lower temperatures at CMN promoting the condensation of semivolatile compounds on the aerosol.

The effect of convective transport of aerosol from low altitudes to CMN in the summer season and its impact on chemical composition is also documented by the measurements performed in July 2008 using the dichotomous sampler. Aerosol collection was concentrated in time of day when upslope breezes are strongest (between 12:00 and 23:00 LT). A parallel set of samples was collected using a second dichotomous sampler deployed at the rural Po Valley station of San Pietro Capofiume (SPC). The results of the ion chromatographic analysis and of evolved gas analysis are reported in Fig. 2. The concentration of reconstructed PM1–10 mass at CMN was about a half that in the Po Valley where significantly higher concentrations of seasalt, soil dust and adsorbed nitrate were observed. By contrast, the total concentration of submicron aerosol compounds in SPC was only 14% higher than at CMN. In particular, the smaller concentrations of sulfate at the mountain site with

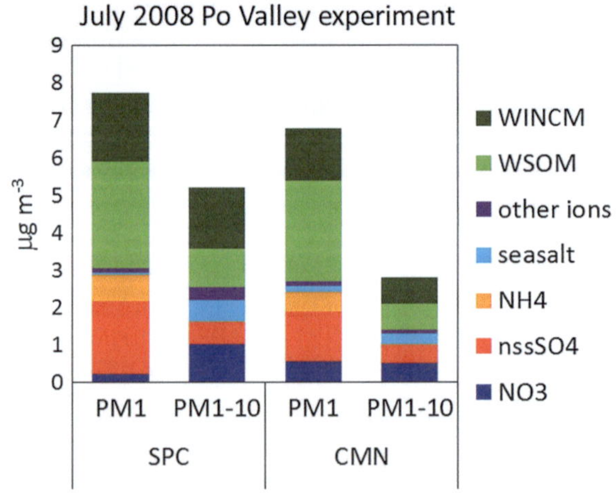

Fig. 2 Concentrations of main carbonaceous and inorganic ionic components of the submicron (PM_1) and supermicron (PM1–10) aerosol fractions in San Pietro Capofiume (SPC) and at CMN during July 2008. The CMN concentrations are reported in standard cubic meters for comparison with the concentrations at SPC which is located at 10 m a.s.l.

respect to the Po Valley are compensated by higher concentrations of submicron nitrate. In agreement with the results of AEROCLOUDS, the higher PM_1 concentrations of the Po Valley are mainly due to water-insoluble carbonaceous matter. This pattern is consistent with the stronger sources of black carbon and primary aerosols from fossil fuel combustion in the Po Valley with respect to CMN. By contrast, water-soluble organic compounds, which are expected to recover secondary organic aerosols, appeared very uniformly distributed between the two sites. The effect of the wind breeze circulation on modifying the PM_1 chemical composition at CMN in summertime is further detailed by the aerosol mass spectrometric (AMS) observations discussed later in this chapter.

4 Desert Dust Outbreaks

Saharan dust plumes transported northward in the free troposphere are responsible for fast, strong increases of PM concentrations at mountain sites on the Alps and Apennines. TSP concentrations of 118 μg m^{-3} have been observed at Jungfraujoch during an outstanding dust event (Schwikowski et al. 1995). Since CMN represents one of the first European mountain ridges that are affected by the transport of mineral dust from Africa, it represents a very suitable location were these events can be observed and their impact to tropospheric composition investigated. Thanks to the continuous observations by an optical particle sizer, Duchi et al. (2016) investigated the long-term variability (2002–2012) of mineral dust occurrence at CMN. By this work, a clear seasonal cycle was evident, with the highest frequency of dust outbreaks from spring to autumn (19.3% of days in summer-spring and 16.3% in autumn), and with a dust-induced variation of the coarse particle number concentration (i.e. particle with diameter larger than 1 μm) higher than 123% on seasonal basis. Dominant mineral dust source regions were recognized on northern western and central Africa. Despite, particle optical and physical properties, the chemical composition of mineral dust particles has not been routinely monitored at CMN, and water-soluble calcium determined by ion chromatography systematically in all field experiments is the only chemical species that can be used to trace the contribution of mineral species (Putaud et al. 2004). Nevertheless, the analysis of insoluble inorganic species was carried out on subsets of samples using X-ray fluorescence (XRF) or particle-induced X-ray emission (Calzolai et al. 2006). The total mineral fraction mass is estimated starting from the measured concentrations of elements assuming simple stoichiometrics for metal and metalloid oxides. Marenco et al. (2006) determined the elemental ratios for the African dust events at CMN: Si/Al = 2.31, Fe/Ca = 0.94, Ca/Al = 0.90, K/Ca = 0.44, Ti/Ca = 0.11, and Ti/Fe = 0.12. Such ratios indicate that African desert particles are enriched in Fe and Ti and depleted in Ca with respect to dust aerosols in background air at CMN. During following experiments, when the mineral fraction was determined on quartz-fiber filters, the Si amount in dust particles was extrapolated from the Al concentrations

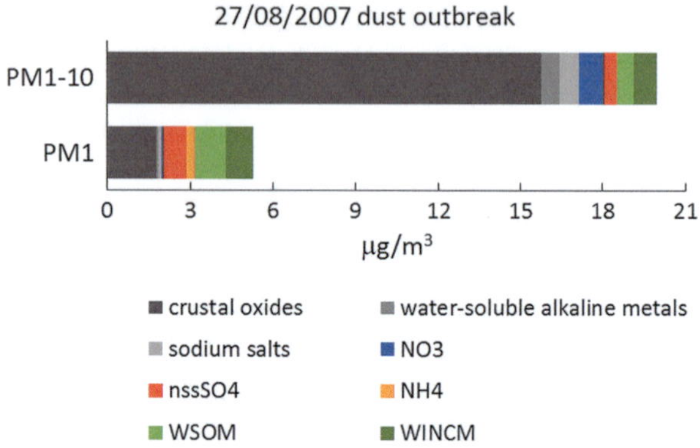

Fig. 3 Concentrations of the mineral fraction, water-soluble inorganic ions and carbonaceous material in the submicron (PM_1) and supermicron (PM1–10) fraction of the aerosol at CMN during the 27 Aug dust outbreak. The mineral dust mass was determined in the supermicron aerosol fraction by particle-induced X-ray emission (PIXE). The ratio between mineral dust and water-soluble calcium was used to estimate the mineral dust mass in PM_1 starting from measured Ca^{2+} analyses

assuming the same Si/Al ratio determined by Marenco et al. (2006) on samples collected on PTFE filters.

The results of the determination of mineral dust mass using chemical analyses provided confirmation of the optical particle counter (OPC) measurements of aerosol volume concentrations indicating that coarse aerosol loadings at CMN during desert dust outbreaks can be massive. During the event of 10 August 2004, the daily concentrations of desert dust peaked at 50 µg m^{-3} (Marenco et al. 2006), while an exceptional Saharan dust event occurred from 13 to 15 March 2004, leading to PM_{10} concentration of about 80 µg m^{-3} (Brattich et al. 2015). Cristofanelli et al. (2009) reported concentrations of coarse-mode mineral dust of the order of 20 µg m^{-3} on a 12-h average basis during the outbreak of 27–30 August 2007 (Fig. 3). The reconstructed supermicron PM mass was more than three times greater than that of PM_1. During a weaker dust outbreak of summer 2000, Putaud et al. (2004) reported a ratio of coarse-to-fine PM of about two. Cristofanelli et al. (2009) showed that ca. 15% of coarse aerosol mass was accounted for organic and inorganic water-soluble species, including nitrate, which was present in much higher concentrations than in the PM_1 fraction. The occurrence of nitrate and sulfate associated to dust aerosols witnesses the production of secondary aerosol mass by heterogeneous reactions of sulfur dioxide and NO_x on the dust particles (Dentener et al. 1996).

5 Volcanic Aerosol Transport

Big volcanic eruptions are well-known modifiers of global climate (IPCC 2013). Ashes and reactive gases can be injected at several kilometers of altitude where they can be transported over long distances and reside for long time in the stratosphere. Volcanic plumes travelling as low as the middle troposphere are more transient in nature, but can pose problems to air traffic as well as deterioration of air quality, as the 2010 eruption of the Eyjafjallajökull has very clearly shown (Gudmundsson et al. 2012). According to historical records, past events of intense volcanism in Iceland posed a serious threat to air quality and agriculture in vast regions in Europe (Thordarson and Self 2003), and the observations performed during the April-Maj 2010 episode indeed documented an impact on atmospheric composition at continental-scale. From the experimental point of view, the breadth of the area affected by the transport of volcanic ashes demonstrated that in situ measurements are feasible not only using research aircrafts but also at stationary observatories, and especially at high-altitude stations. Figure 4 summarizes the concentrations of volcanic aerosols determined from ground-based measurements in central Europe during the two outbreaks occurred on 17–21 April and 18–20 May. Although the volcanic plume was observed as south as Greece and Turkey (Papayannis et al. 2012), the volcanic ashes could cross the Alps with some days of delay during the first phase of the eruption. In this context, the observations performed at ca. 2000 m a.s.l. at CMN were particularly informative, because they provided the concentrations and the characteristics of volcanic aerosols as they entered the Mediterranean basin during their southerly transport. Moreover, thanks to the continuous observation of trace gases and aerosol physical properties, CMN was the only measurement site over Italy able to provide a real-time in-situ update of the volcanic plume evolution in the lower free troposphere (Sandrini et al. 2014). The results of the aerosol physical (number size distribution, absorption/scattering coefficients) and chemical properties observed at CMN during the eruption event are extensively discussed by Sandrini et al. (2014). Briefly, samples of fine (PM_1) and coarse (PM1–10) aerosol were collected using a dichotomous virtual impactor, and the analyses included ion chromatography, OC/EC determination, particle-induced X-ray emission (PIXE), particle-induce gamma ray emission (PIGE), and inductively coupled plasma mass spectroscopy (ICP-MS). These represent unique size-segregated chemical measurements performed on the Eyjafjallajökull aerosols at a European high-altitude site. The analyses allowed to estimate the increase in the concentrations of specific aerosol components during the volcanic dust outbreaks with respect to background air. While volcanic ashes contributed to significant amounts of coarse-mode mineral dust (especially during the second outbreak), the condensation of ammonium sulfate salts (from SO_2 and ammonia emitted by the volcano) contributed to enhance submicron PM concentration. The resulting mass size distribution was largely skewed to the small diameters, with PM_1 accounting for only 30% of total reconstructed PM_{10} mass during the second outbreak. Coarse-mode dust was mainly apportioned to volcanic ashes containing sodium, magnesium and calcium aluminosilicates and titanium and manganese oxides (compatible with the trachyandesite composition of the Eyjafjallajökull rocks). This is in agreement with the observations performed on the northern side of the Alps at Jungfraujoch, where volcanic ashes were observed on both 17–21 April and 18–20 May events.

Coarse-mode volcanic ashes were observed at CMN in much higher amounts during the second outbreak because in that occasion the plume travelled directly from the north Atlantic down to the central Mediterranean (Hervo et al. 2012). By contrast, during the first event, the volcanic ashes did not cross the Alpine barrier

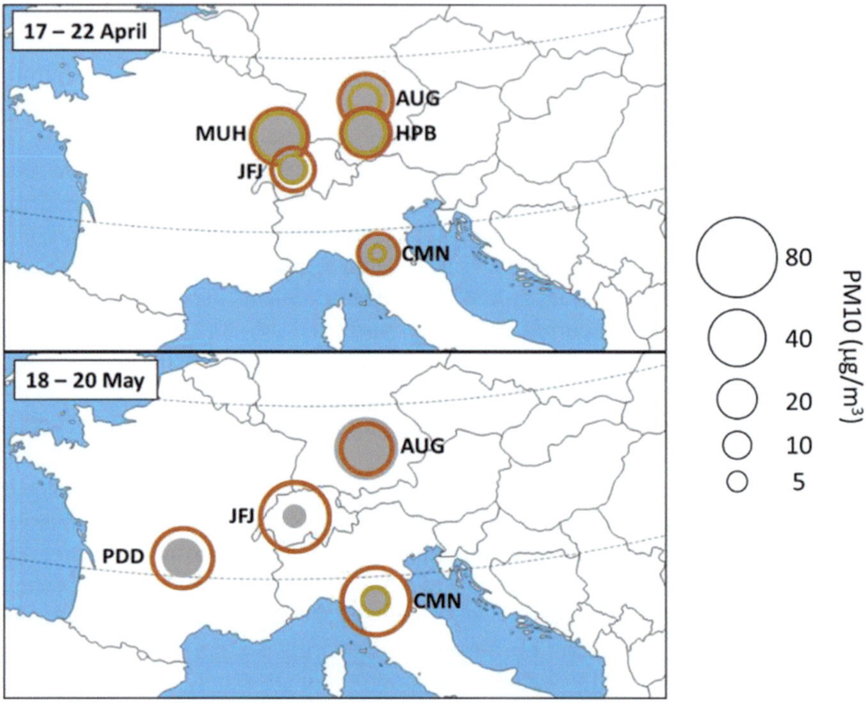

Fig. 4 Concentration of volcanic aerosol at western and central European stations during the first (*top*) and second (*bottom*) outbreak of ashes from the Eyjafjallajökull. *Outer red-brown circles* are scaled to the maximum concentrations of volcanic aerosols determined by high-resolution measurements (like OPC, or scanning mobility particle sizers SMPS), while *inner orange circles* refer to average volcanic aerosol concentrations (including data from analyses of filter samples) and *grey circles* represent the concentrations of background aerosols (PM_{10}). Station codes and references: PDD (Puy-de-Dôme; (Hervo et al. 2012)), Mulhouse (MUH; (Colette et al. 2011)), Jungfraujoch (JFJ; (Bukowiecki et al. 2011)), Hohenpeissenberg (HPB; (Pitz et al. 2011; Schafer et al. 2011)), Augsburg (AUG), Monte Cimone (CMN; (Sandrini et al. 2014))

until 19 April, after 2 days of mixing in the lower troposphere north of the Alps (Bukowiecki et al. 2011; Hervo et al. 2012; Colette et al. 2011; Pitz et al. 2011). Therefore, during the first outbreak, volcanic aerosols reached north Italy and the Mediterranean basin considerably diluted and mixed with background aerosols. Figure 4 shows that, if maximum volcanic aerosols concentrations recorded at CMN were of the same order of magnitude than at northern sites, their average contribution to PM_{10} was significantly smaller than north of the Alps. The different fates of the Eyjafjallajökull plumes of ashes that crossed the European lower troposphere between April and May 2010 were therefore clearly traced by combining the in situ observations at high-altitude sites (Jungfraujoch, Puy-de-Dôme, Monte Cimone), which nicely complemented the measurements performed using research aircrafts and the atmospheric profiling by LiDAR systems (Pappalardo et al. 2013).

6 Aerosol Mass Spectrometric (AMS) Measurements

The online measurements of trace gases and aerosol optical parameters discussed in dedicated chapters of this book clearly show that changes in the weather and in local wind regimes at CMN are responsible for a high degree of short-term variability (1 h or less) in atmospheric composition. Such changes are simply averaged out when sampling aerosol for off-line chemical analysis, as collection time is typically in the range of 12–24 h. The recent development of quantitative aerosol mass spectrometric techniques (especially Aerodyne AMS systems) provided a breakthrough in aerosol observations allowing sensitive, automatic, real-time chemical determinations of non-thermally-refractory aerosol components in PM_1, such as ammonium, nitrate, sulfate, chloride and several organic materials (Jayne et al. 2000; Ng et al. 2011). The first measurements performed at CMN using a high-resolution time-of-flight aerosol mass spectrometer (HR-ToF-AMS) are discussed in the paper by Rinaldi et al. (2015). The observations were carried out in the summer season (June–July 2012) and provided a clear picture of the composition of submicron aerosols components at times of the day when the station resides in the PBL or, alternatively, in the free troposphere. While sulfate is prevalently associated to background aerosols, fine nitrate is clearly transported to the site by upslope breezes flowing from the Po Valley. Finally, the concentrations of organic matter are correlated with specific humidity, indicating that organic particles are enriched in air masses originating from the PBL.

Similar observations were carried at other high-altitude research stations in central-western Europe. Figure 5 provides a focus on experiments carried out in the warm season. The relatively small concentrations at the site located at lower altitude

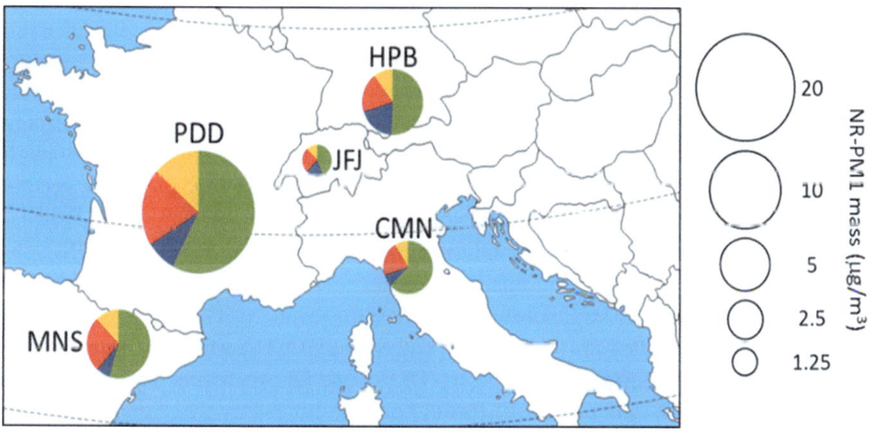

Fig. 5 Concentrations of non-refractory submicron aerosol components determined by aerosol mass spectrometric (AMS) measurements at high-altitude sites in Europe between spring and summer months. Chemical components and color code: nitrate (*blue*), sulfate (*red*), ammonium (*orange*), organic matter (*green*). Station codes and references: Puy-de-Dôme (PDD; (Freney et al. 2011)), Jungfraujoch (JFJ; (Lanz et al. 2010)), Hohenpeissenberg (HPB; (Hock et al. 2008)), Monte Cimone (CMN; (Rinaldi et al. 2015)), Montsec (MNS; Ripoll et al. 2015)

(Hohenpeissenberg, 985 m a.s.l.) are probably due the early period of the year when the observations were performed (spring instead of summer) (Hock et al. 2008). The concentrations at the highest station (Jungfraujoch, 3571 m a.s.l.) are by far the smallest and characterized by a relative large amount of nitrate (Lanz et al. 2010). The smaller nitrate contributions at CMN and Montsec (1570 m a.s.l.; (Ripoll et al. 2015)) are probably due to the higher temperatures leading to the volatilization of nitric acid. Organic matter is the major PM_1 component measured by AMS at all sites, but its contribution is the highest at CMN (63%). More information on the origin of particulate organic compounds at CMN is provided in the following discussion.

7 Origin of Organic Matter

Organic matter represents the main constituent of submicron aerosol mass at CMN and its contribution has even increased since the early 2000s as a consequence of the decline of sulfate concentrations. The lack of long-term negative trends for organic matter in spite of the historical decrease of anthropogenic emissions in Italy and Europe cannot be easily explained, given the great diversity of sources and processes contributing to organic aerosol formation at large spatial scales (Kanakidou et al. 2005). Source apportionment of aerosol organic matter is particularly challenging at remote sites where the chemical composition of organic particles is highly modified with respect to the emissions and molecular tracers are often undetectable. Spectral deconvolution of the timeline of electron impact mass spectra provided by the HR-ToF-AMS measurements at CMN in summer 2012 identified diverse organic source-contributions but none of them could be attributed unambiguously to primary combustion sources such as traffic or biomass burning (Rinaldi et al. 2015). The three organic contributions disentangled by positive matrix factorization (PMF) all showed spectral profiles typical of oxidized organic aerosols (OOAs) which are usually associated to secondary organic aerosol (SOA) (Zhang et al. 2007). The results for CMN, like for other background stations (Freney et al. 2011), show that the spectral fingerprints of the specific OOA types reflect more the ageing state of SOA rather than actual source fingerprints. In particular, increasing oxidation states were observed for OOA compounds (1) originating from PBL air transported to the Apennines by upslope breezes, (2) transported in residual layers (recirculated PBL air), (3) transported in the free troposphere. The time trends of the first and of the third class correlated well with those of CO and ozone, respectively (Fig. 6). The first and third OOA fractions (OOAa and OOAc, respectively) occurred at relatively constant concentrations throughout the June–July 2012 and their relative contributions followed the consistent diurnal variability of the mountain breeze system. By contrast, the second class, OOAb, – showing an intermediate oxidation state – occurred in significant amounts only under stagnation periods when it could account for 80% of OM, while its contribution was only 25% on a campaign average. The first class, OOAa, tracing the contribution of sources in the Po Valley and

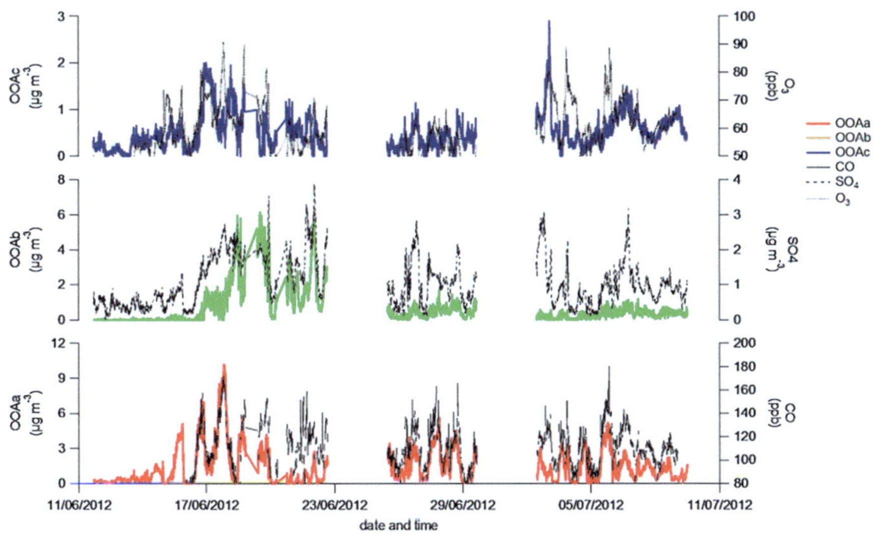

Fig. 6 Concentration time trends for the three fractions (OOAa, b and c) of aerosol organic matter determined by PMF-AMS during the summer 2012 field campaign at CMN Rinaldi et al. (2015)

in the north Apennines, accounted on average for more than a half (55%) of organic mass. The remaining 20% (OOAc) must be attributed to prolonged atmospheric processing of organic aerosol occurring mainly in the free troposphere. Therefore, if the analysis of AMS spectra could not identify the actual biogenic vs. anthropogenic sources of organic mass at CMN, important information could be gained on their main spatial footprints.

The investigation of the anthropogenic and biogenic sources of OOAs at CMN is still subject of investigation and analytical techniques other than AMS have been employed. In particular, nuclear magnetic resonance (NMR) spectroscopy can be profitably be exploited for source attribution of oxidized, water-soluble aerosol organic compounds (Decesari et al. 2007; Paglione et al. 2014). Figure 7 shows examples of 400 MHz proton NMR spectra for WSOC PM_{10} samples collected at CMN in summer and winter. The summer sample, which was collected in a period of prolonged stagnant atmospheric conditions in 2003, shows a complexity of resonances in the aliphatic part of the spectrum, especially in the regions of alkyls (C-H) and acyls (H-C-C = O) regions. Water-soluble organic aerosols with analogous NMR spectral fingerprint were recently reproduced in laboratory starting from biogenic (monoterpene) SOA (Zanca et al. 2017). The winter sample shows very different spectral features. The occurrence of aromatic compounds and of traces of levoglucosan, the best known tracer of cellulose pyrolysis, indicate that biomass burning sources can contribute to organic aerosols concentrations even in the free troposphere at this site. Spectral fingerprints characteristic of aged biomass burning products were found also in summertime at CMN during episodes of transport of smoke plumes from big wildfires (Cristofanelli et al. 2009; Paglione et al. 2014). All

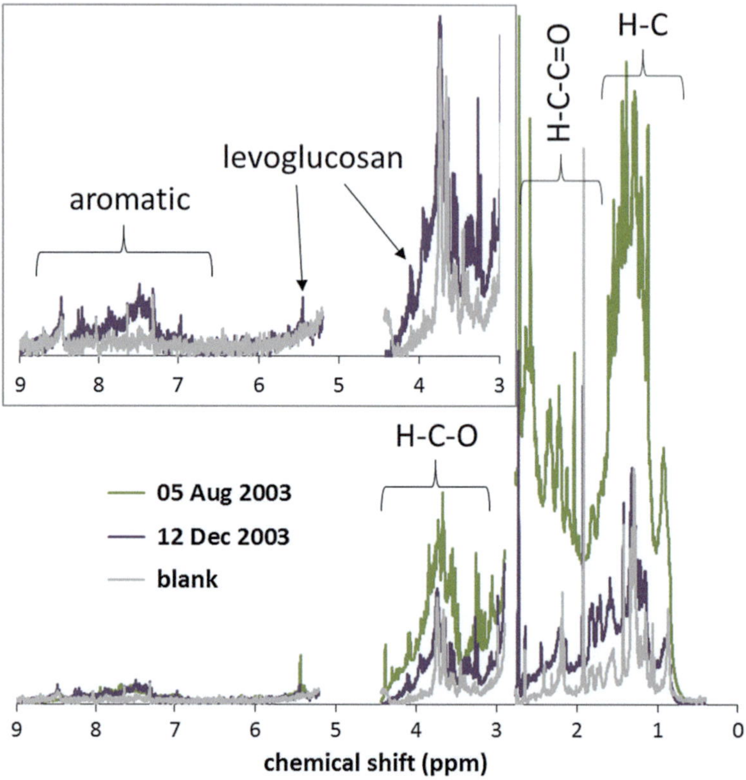

Fig. 7 Proton nuclear magnetic resonance (NMR) spectra of WSOC extracted from two PM_{10} samples from CMN. The analyzed functional groups include: (**a**) aromatics, (**b**) alcoxyl (H-C-O), (**c**) acyl (H-C-C = O) and (**d**) alkyl (C-H). The insert reports a zoom of the winter sample in the spectral region between 3 and 9 ppm of chemical shift

the NMR spectral data recorded for CMN samples are specific for WSOC. The biogenic/anthropogenic nature of water-insoluble organic matter at CMN remains unknown.

Acknowledgments The author would like to acknowledge the full team of scientists and technicians of the Institute of Atmospheric Sciences and Climate (ISAC) involved in the activities of aerosol sampling in Monte Cimone and in the laboratory chemical analysis. In particular, for the field work at CMN: A. Marinoni, U. Bonafé, F. Calzolari, R. Duchi. For the AMS measurements: M. Rinaldi and S. Gilardoni. For the laboratory workup and analytical chemistry: E. Matta, L. Emblico, E. Finessi, L. Giulianelli, and two researchers who extensively examined the Mt. Cimone chemical datasets, C. Carbone, S. Sandrini, that I greatly acknowledge. Finally, the author would like to thank the PIs of the atmospheric chemistry group of ISAC in Bologna, M. C. Facchini and S. Fuzzi.

References

Bourcier L, Sellegri K, Chausse P, Pichon JM, Laj P (2012) Seasonal variation of water-soluble inorganic components in aerosol size-segregated at the puy de Dome station (1,465 m a.s.l.), France. J Atmos Chem 69:47–66. doi:10.1007/s10874-012-9229-2

Brattich E, Hernández-Ceballos M, Cinelli G, Tositti L (2015) Analysis of 210Pb peak values at Mt. Cimone (1998–2011). Atmos Environ 112:136–147

Bukowiecki N, Zieger P, Weingartner E, Juranyi Z, Gysel M, Neininger B, Schneider B, Hueglin C, Ulrich A, Wichser A, Henne S, Brunner D, Kaegi R, Schwikowski M, Tobler L, Wienhold FG, Engel I, Buchmann B, Peter T, Baltensperger U (2011) Ground-based and airborne in-situ measurements of the Eyjafjallajokull volcanic aerosol plume in Switzerland in spring 2010. Atmos Chem Phys 11:10011–10030. doi:10.5194/acp-11-10011-2011

Calzolai G, Chiari M, Orellana IG, Lucarelli F, Migliori A, Nava S, Taccetti F (2006) The new external beam facility for environmental studies at the Tandetron accelerator of LABEC. Nucl Instrum Method Phys Res Sect B-Beam Interact Mater Atoms 249:928–931. doi:10.1016/j.nimb.2006.03.193

Carbone C, Decesari S, Mircea M, Giulianelli L, Finessi E, Rinaldi M, Fuzzi S, Marinoni A, Duchi R, Perrino C, Sargolini T, Varde M, Sprovieri F, Gobbi GP, Angelini F, Facchini MC (2010) Size-resolved aerosol chemical composition over the Italian Peninsula during typical summer and winter conditions. Atmos Environ 44:5269–5278. doi:10.1016/j.atmosenv.2010.08.008

Carbone C, Decesari S, Paglione M, Giulianelli L, Rinaldi M, Marinoni A, Cristofanelli P, Didiodato A, Bonasoni P, Fuzzi S, Facchini MC (2014) 3-year chemical composition of free tropospheric PM1 at the Mt. Cimone GAW global station – South Europe – 2165 m a.s.l. Atmos Environ 87:218–227. doi:10.1016/j.atmosenv.2014.01.048

Cavalli F, Alastuey A, Areskoug H, Ceburnis D, Cech J, Genberg J, Harrison RM, Jaffrezo JL, Kiss G, Laj P, Mihalopoulos N, Perez N, Quincey P, Schwarz J, Sellegri K, Spindler G, Swietlicki E, Theodosi C, Yttri KE, Aas W, Putaud JP (2016) A European aerosol phenomenology-4: harmonized concentrations of carbonaceous aerosol at 10 regional background sites across Europe. Atmos Environ 144:133–145. doi:10.1016/j.atmosenv.2012.07.050

Colette A, Favez O, Meleux F, Chiappini L, Haeffelin M, Morille Y, Malherbe L, Papin A, Bessagnet B, Menut L, Leoz E, Rouil L (2011) Assessing in near real time the impact of the April 2010 Eyjafjallajokull ash plume on air quality. Atmos Environ 45:1217–1221. doi:10.1016/j.atmosenv.2010.09.064

Cozic J, Verheggen B, Weingartner E, Crosier J, Bower KN, Flynn M, Coe H, Henning S, Steinbacher M, Henne S, Coen MC, Petzold A, Baltensperger U (2008) Chemical composition of free tropospheric aerosol for PM1 and coarse mode at the high alpine site Jungfraujoch. Atmos Chem Phys 8:407–423

Cristofanelli P, Marinoni A, Arduini J, Bonafe U, Calzolari F, Colombo T, Decesari S, Duchi R, Facchini MC, Fierli F, Finessi E, Maione M, Chiari M, Calzolai G, Messina P, Orlandi E, Roccato F, Bonasoni P (2009) Significant variations of trace gas composition and aerosol properties at Mt. Cimone during air mass transport from North Africa – contributions from wildfire emissions and mineral dust. Atmos Chem Phys 9:4603–4619

Decesari S, Mircea M, Cavalli F, Fuzzi S, Moretti F, Tagliavini E, Facchini MC (2007) Source attribution of water-soluble organic aerosol by nuclear magnetic resonance spectroscopy. Environ Sci Technol 41:2479–2484. doi:10.1021/es0617111

Dentener FJ, Carmichael GR, Zhang Y, Lelieveld J, Crutzen PJ (1996) Role of mineral aerosol as a reactive surface in the global troposphere. J Geophys Res-Atmos 101:22869–22889. doi:10.1029/96jd01818

Duchi R, Cristofanelli P, Landi TC et al (2016) Long-term (2002–2012) investigation of Saharan dust transport events at Mt. Cimone GAW global station, Italy (2165 m a.s.l.) Elem Sci Anth. doi:10.12952/journal.elementa.000085

Freney EJ, Sellegri K, Canonaco F, Boulon J, Hervo M, Weigel R, Pichon JM, Colomb A, Prevot ASH, Laj P (2011) Seasonal variations in aerosol particle composition at the puy-de-Dome research station in France. Atmos Chem Phys 11:13047–13059. doi:10.5194/acp-11-13047-2011

Gudmundsson MT, Thordarson T, Hoskuldsson A, Larsen G, Bjornsson H, Prata FJ, Oddsson B, Magnusson E, Hognadottir T, Petersen GN, Hayward CL, Stevenson JA, Jonsdottir I (2012) Ash generation and distribution from the April-May 2010 eruption of Eyjafjallajokull, Iceland. Sci Rep 2. doi:10.1038/srep00572

Henning S, Weingartner E, Schwikowski M, Gaggeler HW, Gehrig R, Hinz KP, Trimborn A, Spengler B, Baltensperger U (2003) Seasonal variation of water-soluble ions of the aerosol at the high-alpine site Jungfraujoch (3580 m asl). J Geophys Res-Atmos 108. doi:10.1029/2002jd002439

Hervo M, Quennehen B, Kristiansen NI, Boulon J, Stohl A, Freville P, Pichon JM, Picard D, Labazuy P, Gouhier M, Roger JC, Colomb A, Schwarzenboeck A, Sellegri K (2012) Physical and optical properties of 2010 Eyjafjallajokull volcanic eruption aerosol: ground-based, Lidar and airborne measurements in France. Atmos Chem Phys 12:1721–1736. doi:10.5194/acp-12-1721-2012

Hock N, Schneider J, Borrmann S, Rompp A, Moortgat G, Franze T, Schauer C, Poschl U, Plass-Dulmer C, Berresheim H (2008) Rural continental aerosol properties and processes observed during the Hohenpeissenberg aerosol characterization experiment (HAZE2002). Atmos Chem Phys 8:603–623

Husain L, Parekh PP, Dutkiewicz VA, Khan AR, Yang K, Swami K (2004) Long-term trends in atmospheric concentrations of sulfate, total sulfur, and trace elements in the northeastern United States. J Geophys Res-Atmos 109. doi:10.1029/2004jd004877

IPCC (2013) Climate change 2013: the physical science basis. Contribution of working group I to the fifth assessment report of the Intergovermental Panel on Climate Change. In: Stocker TF, Qin D, Plattner G-K, Tignor M, Allen SK, Boschung J, Nauels A, Xia Y, Bex V, Midgley PM (eds) . Cambridge University Press, Cambridge/New York

Jayne JT, Leard DC, Zhang XF, Davidovits P, Smith KA, Kolb CE, Worsnop DR (2000) Development of an aerosol mass spectrometer for size and composition analysis of submicron particles. Aerosol Sci Technol 33:49–70. doi:10.1080/027868200410840

Kanakidou M, Seinfeld JH, Pandis SN, Barnes I, Dentener FJ, Facchini MC, Van Dingenen R, Ervens B, Nenes A, Nielsen CJ, Swietlicki E, Putaud JP, Balkanski Y, Fuzzi S, Horth J, Moortgat GK, Winterhalter R, Myhre CEL, Tsigaridis K, Vignati E, Stephanou EG, Wilson J (2005) Organic aerosol and global climate modelling: a review. Atmos Chem Phys 5:1053–1123

Kasper A, Puxbaum H (1998) Seasonal variation of SO2, HNO3, NH3 and selected aerosol components at Sonnblick (3106 m asl). Atmos Environ 32:3925–3939. doi:10.1016/s1352-2310(97)00031-9

Lanz VA, Prevot ASH, Alfarra MR, Weimer S, Mohr C, DeCarlo PF, Gianini MFD, Hueglin C, Schneider J, Favez O, D'Anna B, George C, Baltensperger U (2010) Characterization of aerosol chemical composition with aerosol mass spectrometry in Central Europe: an overview. Atmos Chem Phys 10:10453–10471. doi:10.5194/acp-10-10453-2010

Marenco F, Bonasoni P, Calzolari F, Ceriani M, Chiari M, Cristofanelli P, D'Alessandro A, Fermo P, Lucarelli F, Mazzei F, Nava S, Piazzalunga A, Prati P, Valli G, Vecchi R (2006) Characterization of atmospheric aerosols at Monte Cimone, Italy, during summer 2004: source apportionment and transport mechanisms. J Geophys Res-Atmos 111. doi:10.1029/2006jd007145

Matta E, Facchini MC, Decesari S, Mircea M, Cavalli F, Fuzzi S, Putaud JP, Dell'Acqua A (2003) Mass closure on the chemical species in size-segregated atmospheric aerosol collected in an urban area of the Po Valley, Italy. Atmos Chem Phys 3:623–637

Ng NL, Herndon SC, Trimborn A, Canagaratna MR, Croteau PL, Onasch TB, Sueper D, Worsnop DR, Zhang Q, Sun YL, Jayne JT (2011) An Aerosol Chemical Speciation Monitor (ACSM) for routine monitoring of the composition and mass concentrations of ambient aerosol. Aerosol Sci Technol 45:780–794. doi:10.1080/02786826.2011.560211

Paglione M, Saarikoski S, Carbone S, Hillamo R, Facchini MC, Finessi E, Giulianelli L, Carbone C, Fuzzi S, Moretti F, Tagliavini E, Swietlicki E, Stenstrom KE, Prevot ASH, Massoli P,

Canaragatna M, Worsnop D, Decesari S (2014) Primary and secondary biomass burning aerosols determined by proton nuclear magnetic resonance (H-1-NMR) spectroscopy during the 2008 EUCAARI campaign in the Po Valley (Italy). Atmos Chem Phys 14:5089–5110. doi:10.5194/acp-14-5089-2014

Papayannis A, Mamouri RE, Amiridis V, Giannakaki E, Veselovskii I, Kokkalis P, Tsaknakis G, Balis D, Kristiansen NI, Stohl A, Korenskiy M, Allakhverdiev K, Huseyinoglu MP, Baykara T (2012) Optical properties and vertical extension of aged ash layers over the Eastern Mediterranean as observed by Raman lidars during the Eyjafjallajokull eruption in may 2010. Atmos Environ 48:56–65. doi:10.1016/j.atmosenv.2011.08.037

Pappalardo G, Mona L, D'Amico G, Wandinger U, Adam M, Amodeo A, Ansmann A, Apituley A, Arboledas LA, Balis D, Boselli A, Bravo-Aranda JA, Chaikovsky A, Comeron A, Cuesta J, De Tomasi F, Freudenthaler V, Gausa M, Giannakaki E, Giehl H, Giunta A, Grigorov I, Gross S, Haeffelin M, Hiebsch A, Iarlori M, Lange D, Linne H, Madonna F, Mattis I, Mamouri RE, McAuliffe MAP, Mitev V, Molero F, Navas-Guzman F, Nicolae D, Papayannis A, Perrone MR, Pietras C, Pietruczuk A, Pisani G, Preissler J, Pujadas M, Rizi V, Ruth AA, Schmidt J, Schnell F, Seifert P, Serikov I, Sicard M, Simeonov V, Spinelli N, Stebel K, Tesche M, Trickl T, Wang X, Wagner F, Wiegner M, Wilson KM (2013) Four-dimensional distribution of the 2010 Eyjafjallajokull volcanic cloud over Europe observed by EARLINET. Atmos Chem Phys 13:4429–4450. doi:10.5194/acp-13-4429-2013

Pio CA, Legrand M, Oliveira T, Afonso J, Santos C, Caseiro A, Fialho P, Barata F, Puxbaum H, Sanchez-Ochoa A, Kasper-Giebl A, Gelencser A, Preunkert S, Schock M (2007) Climatology of aerosol composition (organic versus inorganic) at nonurban sites on a west-east transect across Europe. J Geophys Res-Atmos 112. doi:10.1029/2006jd008038

Pitz M, Gu J, Soentgen J, Peters A, Cyrys J (2011) Particle size distribution factor as an indicator for the impact of the Eyjafjallajokull ash plume at ground level in Augsburg, Germany. Atmos Chem Phys 11:9367–9374. doi:10.5194/acp-11-9367-2011

Putaud JP, Van Dingenen R, Dell'Acqua A, Raes F, Matta E, Decesari S, Facchini MC, Fuzzi S (2004) Size-segregated aerosol mass closure and chemical composition in Monte Cimone (I) during MINATROC. Atmos Chem Phys 4:889–902

Rinaldi M, Gilardoni S, Paglione M, Sandrini S, Fuzzi S, Massoli P, Bonasoni P, Cristofanelli P, Marinoni A, Poluzzi V, Decesari S (2015) Organic aerosol evolution and transport observed at Mt. Cimone (2165 m a.s.l.), Italy, during the PEGASOS campaign. Atmos Chem Phys 15:11327–11340. doi:10.5194/acp-15-11327-2015

Ripoll A, Minguillon MC, Pey J, Jimenez JL, Day DA, Sosedova Y, Canonaco F, Prevot ASH, Querol X, Alastuey A (2015) Long-term real-time chemical characterization of submicron aerosols at Montsec (southern Pyrenees, 1570 m a.s.l.) Atmos Chem Phys 15:2935–2951. doi:10.5194/acp-15-2935-2015

Sandrini S, Giulianelli L, Decesari S, Fuzzi S, Cristofanelli P, Marinoni A, Bonasoni P, Chiari M, Calzolai G, Canepari S, Perrino C, Facchini MC (2014) In situ physical and chemical characterisation of the Eyjafjallajokull aerosol plume in the free troposphere over Italy. Atmos Chem Phys 14:1075–1092. doi:10.5194/acp-14-1075-2014

Schafer K, Thomas W, Peters A, Ries L, Obleitner F, Schnelle-Kreis J, Birmili W, Diemer J, Fricke W, Junkermann W, Pitz M, Emeis S, Forkel R, Suppan P, Flentje H, Gilge S, Wichmann HE, Meinhardt F, Zimmermann R, Weinhold K, Soentgen J, Munkel C, Freuer C, Cyrys J (2011) Influences of the 2010 Eyjafjallajokull volcanic plume on air quality in the northern Alpine region. Atmos Chem Phys 11:8555–8575. doi:10.5194/acp-11-8555-2011

Schwikowski M, Seibert P, Baltensperger U, Gaggeler HW (1995) A study of an outstanding Saharan dust event at the high-alpine site Jungfraujoch, Switzerland. Atmos Environ 29:1829–1842. doi:10.1016/1352-2310(95)00060-c

Thordarson T, Self S (2003) Atmospheric and environmental effects of the 1783–1784 Laki eruption: a review and reassessment. J Geophys Res-Atmos 108. doi:10.1029/2001jd002042

Zanca N, Lambe AT, Massoli P, Paglione M, Croasdale DR, Parmar Y, Tagliavini E, Gilardoni S, Decesari S (2017) Characterizing source fingerprints and ageing processes in laboratory-generated secondary organic aerosols using proton-nuclear magnetic resonance (1^H–NMR) analysis and HPLC HULIS determination, Atmospheric Chemistry and Physics, 10.5194/acp-2017-91

Zhang Q, Jimenez JL, Canagaratna MR, Allan JD, Coe H, Ulbrich I, Alfarra MR, Takami A, Middlebrook AM, Sun YL, Dzepina K, Dunlea E, Docherty K, DeCarlo PF, Salcedo D, Onasch T, Jayne JT, Miyoshi T, Shimono A, Hatakeyama S, Takegawa N, Kondo Y, Schneider J, Drewnick F, Borrmann S, Weimer S, Demerjian K, Williams P, Bower K, Bahreini R, Cottrell L, Griffin RJ, Rautiainen J, Sun JY, Zhang YM, Worsnop DR (2007) Ubiquity and dominance of oxygenated species in organic aerosols in anthropogenically-influenced Northern Hemisphere midlatitudes. Geophys Res Lett 34. doi:10.1029/2007gl029979

Statistical Analysis of Essential Climate Variables (ECVs) at Mt. Cimone

1 Introduction

This appendix presents the climatological estimates calculated over the two last decades (1996–2015) of essential climate variables (ECVs) collected at the "O. Vittori" observatory, i.e.: air temperature, air pressure, wind speed, relative humidity, surface ozone (O_3), equivalent black carbon (BC), accumulation and coarse particles, methane (CH_4), nitrous oxide (N_2O), sulfur hexafluoride (SF_6), 1,1,1,2-Tetrafluoroethane (HFC-134a), Trichlorofluoromethane (CFC-11), and 1,2-Dichlorotetrafluoroethane (CFC-114). For each atmospheric component, we present the annual variation (upper panel in each figure), the average seasonal diurnal variation (middle panel) and the trend estimate over the entire dataset (bottom panel). Moreover, Table 1 reports the main statistical values for each parameter.

2 Dataset and Methods

The dataset used for this appendix has been made available through the MOVIDA – Monte Cimone On-line Visualization and Data Analyses tool (http://www.isac.cnr.it/cimone/data-access), developed by CNR-ISAC in collaboration with Arpae-SIMC Emilia Romagna. Hourly average values for the different ECVs were used to calculate monthly statistics, typical seasonal mean behavior and long term tendency of monthly averaged values. The MOVIDA data are the same that are annually incorporated into the WMO/GAW World Data Center for Greenhouse Gases (WDCGG), Reactive Gases (WDCRG) and Aerosol (WDCA) hosted by the Japan Meteorological Agency (http://ds.data.jma.go.jp/gmd/wdcgg/) and EBAS system (http://ebas.nilu.no/). These data centers host observation data of atmospheric chemical composition and physical properties coming from national and international programs, ranging from monitoring activities to research projects.

P. Cristofanelli et al., *High-Mountain Atmospheric Research*, SpringerBriefs in Meteorology, DOI 10.1007/978-3-319-61127-3_6

Table 1 Main statistical values for the major atmospheric parameters collected at the GAW global station "O. Vittori" of Mt. Cimone

Parameter (unit)	Min	1st qu.	Median	Mean	3rd qu.	Max	Period
Air temperature (°C)	−21.2	−1.6	2.9	3.2	8.3	8.3	1996–2015
Air pressure (hPa)	751	785	790	789	794	805	1996–2015
Wind speed (m/s)	0.0	4.0	6.7	7.9	10.8	53.2	1996–2015
Relative humidity (%)	0.2	64.7	87.0	78.9	98.8	100.0	1996–2015
O_3 (ppb)	8.9	45.5	52.5	53.5	60.6	218.3	1996–2015
BC (μg/m^3)	UDL	0.2	0.7	1.3	1.7	28.7	2005–2015
Accumulation particle number (cm^{-3})	UDL	2.8	11.2	23.5	29.8	448.8	2002–2015
Coarse particle number (cm^{-3})	UDL	0.02	0.09	0.22	0.23	23.46	2002–2015
CH_4 (ppb)	1793	1862	1881	1887	1905	2270	2008–2015
N_2O (ppb)	320	324	325	326	327	338	2008–2015
SF_6 (ppt)	6.0	7.3	7.8	7.9	8.5	11.9	2009–2015
HFC-134a (ppt)	18.0	49.9	63.3	62.6	75.0	445.2	2001–2015
CFC-11 (ppt)	230.7	239.4	243.1	245.9	252.6	292.2	2002–2015
CFC-114 (ppt)	15.9	16.3	16.4	16.4	16.5	17.2	2006–2015

UDL refers to "under detection limit" given by each instrument

Annual variations were reported by calculating (over the period of data availability of each ECV, see Table 1) the average monthly mean value, as well as the 25th, 50th and 75th percentiles. The average seasonal diurnal variations were obtained by averaging, for each hour (local time: UTC + 1) and season (winter: DJF, spring: MAM, summer: JJA, autumn: SON), the hourly mean values available by the MOVIDA system.

For investigating the long-term trends of the several parameters, we adopted the Theil-Sen method (Theil 1950; Sen 1968) implemented in the Openair software (Carslaw and Ropkins 2012). The basic idea is as follows: given a set of n (x, y) pairs of points, the slopes between all pairs are calculated. Note that the number of slopes can increase by $\approx n^2$ so that it can grow rapidly as the length of the data set increases. The computed Theil-Sen estimate of the slope is the median of all these slopes.

The advantage of using the Theil-Sen estimator, with respect to other methodologies, is that it tends to yield accurate confidence intervals even with non-normal data non-constant error variance. It is also resistant to outliers, which is an important feature for methodologies devoted to the analysis of time series of atmospheric compounds. As previously mentioned, the estimates of these parameters can be made more robust through bootstrap-resampling, which further adds to the computational burden, but is not an issue for most time series, which are expressed either as monthly or annual means. Bootstrap resampling also provides the estimate of p for the slope. For more details see Carslaw and Ropkins (2012).

The Theil-Sen method was applied to the deseasonalised monthly mean values of the ECVs observed at CMN. Only months with at least 75% of hourly data available, were taken into account for the analysis.

Fig. 1 (continued) shows the average seasonal diurnal variations. In the *bottom* panel, the overall trend is presented, along with the 95% confidence intervals in the slope. The trend significance is indicated as follows: $p < 0.001$ = ****, $p < 0.01$ = **, $p < 0.05$ = * and $p < 0.1$ = +

2.1 Air Temperature (Fig. 1)

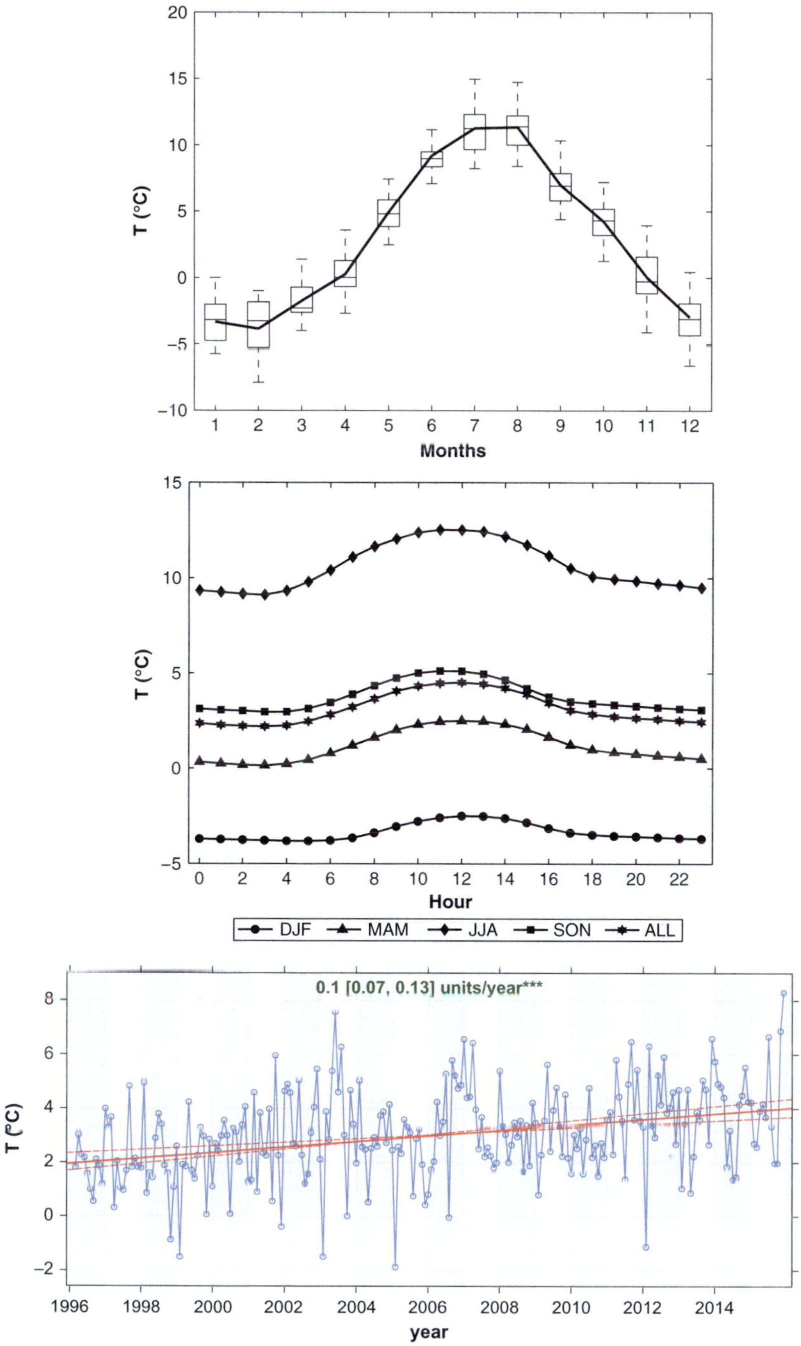

Fig. 1 Statistical analysis of air temperature (T) at CMN, as described in Sect. 1. The upper plot reports the averaged seasonal cycle (*continuous line*) and monthly 25th, 50th and 75th percentiles; the whiskers extend to the most extreme hourly values not considered outliers. The *middle* panel

2.2 *Atmospheric Pressure (Fig. 2)*

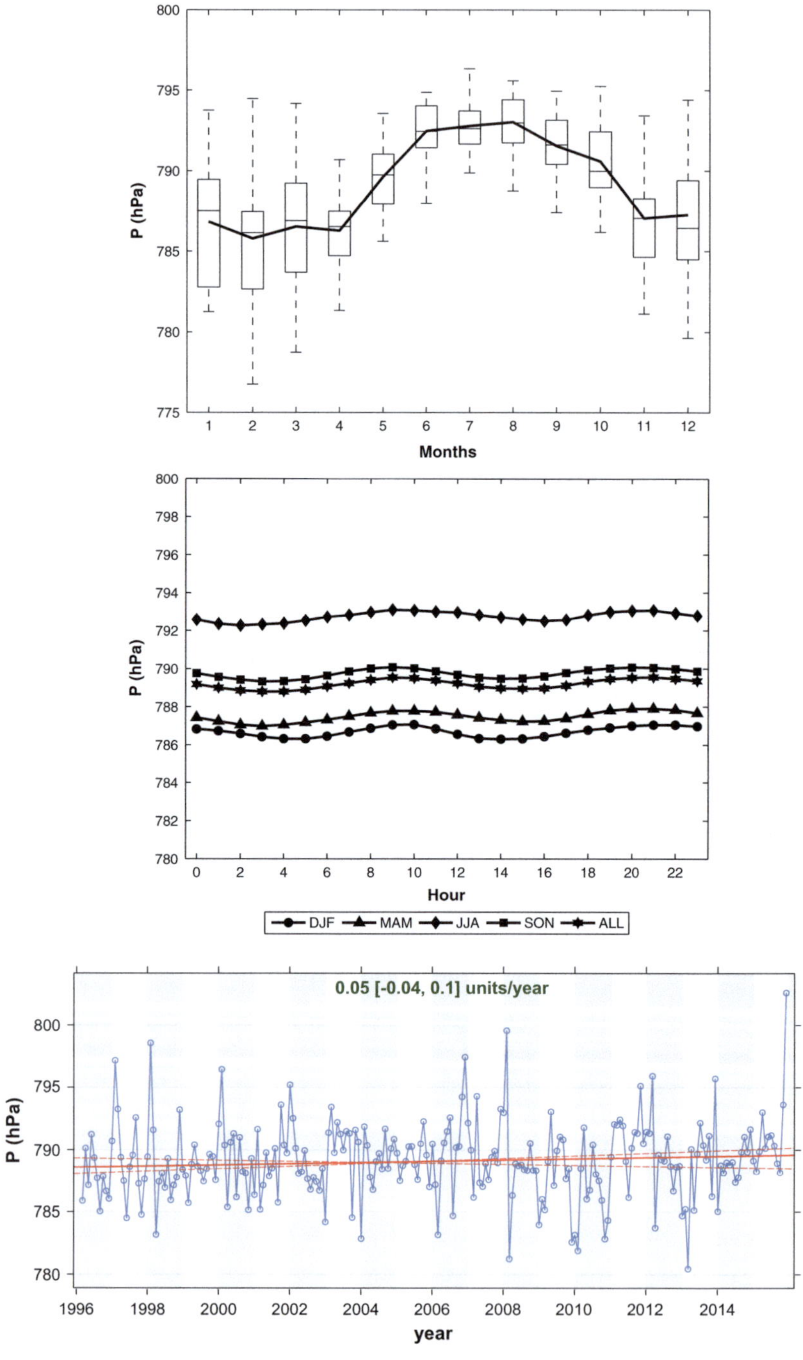

Fig. 2 Same as Fig. 1, for air pressure (P)

2.3 Wind Speed (Fig. 3)

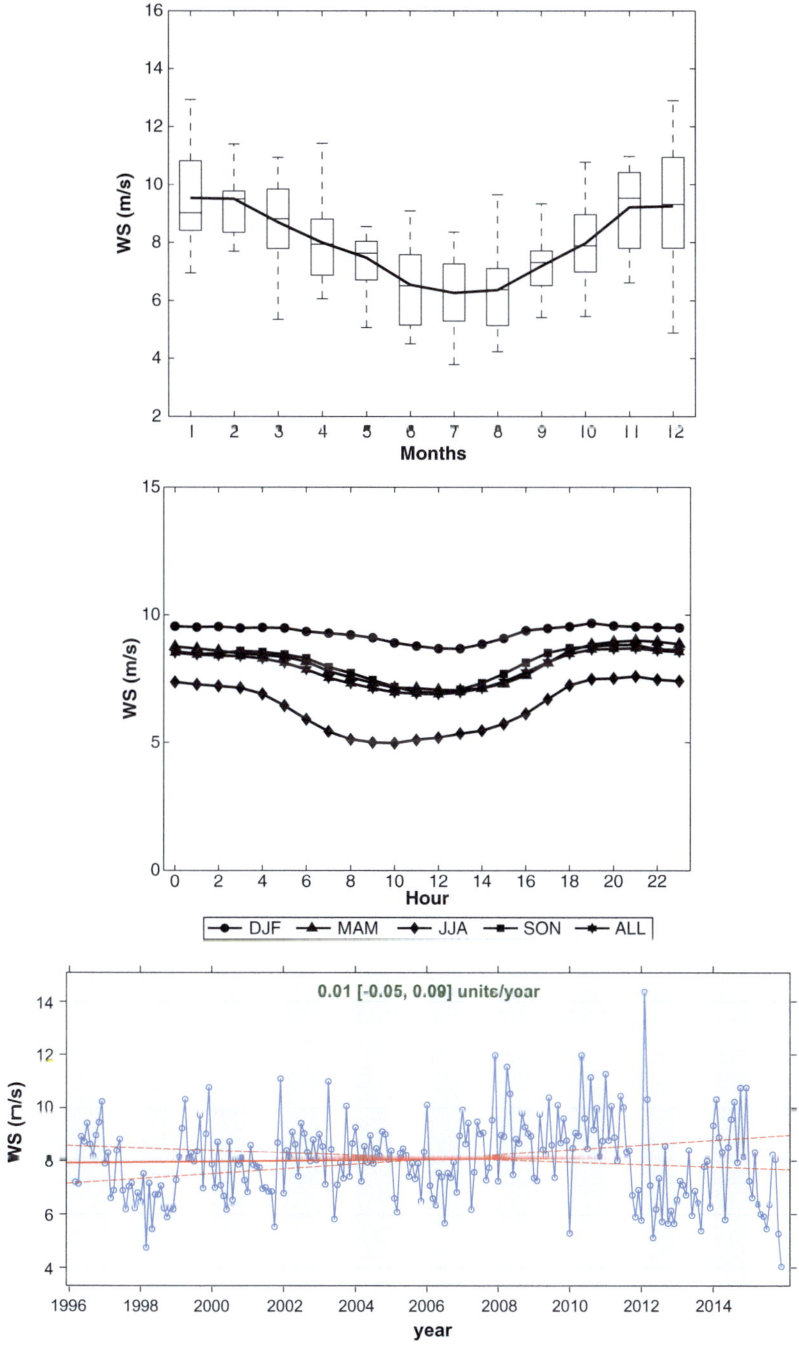

Fig. 3 Same as Fig. 1, for wind speed (WS)

2.4 Relative Humidity (Fig. 4)

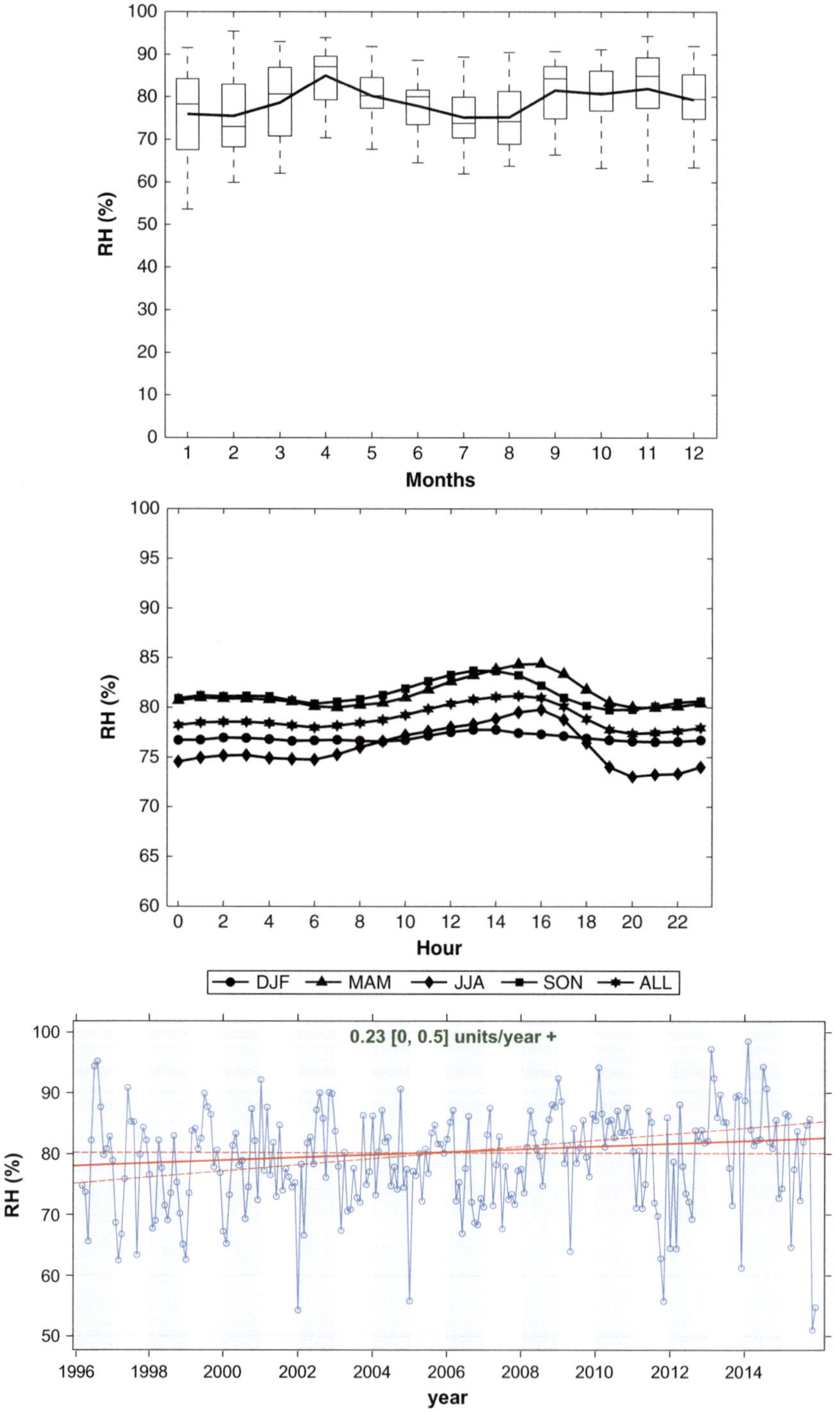

Fig. 4 Same as Fig. 1, for relative humidity (RH)

2.5 Ozone (Fig. 5)

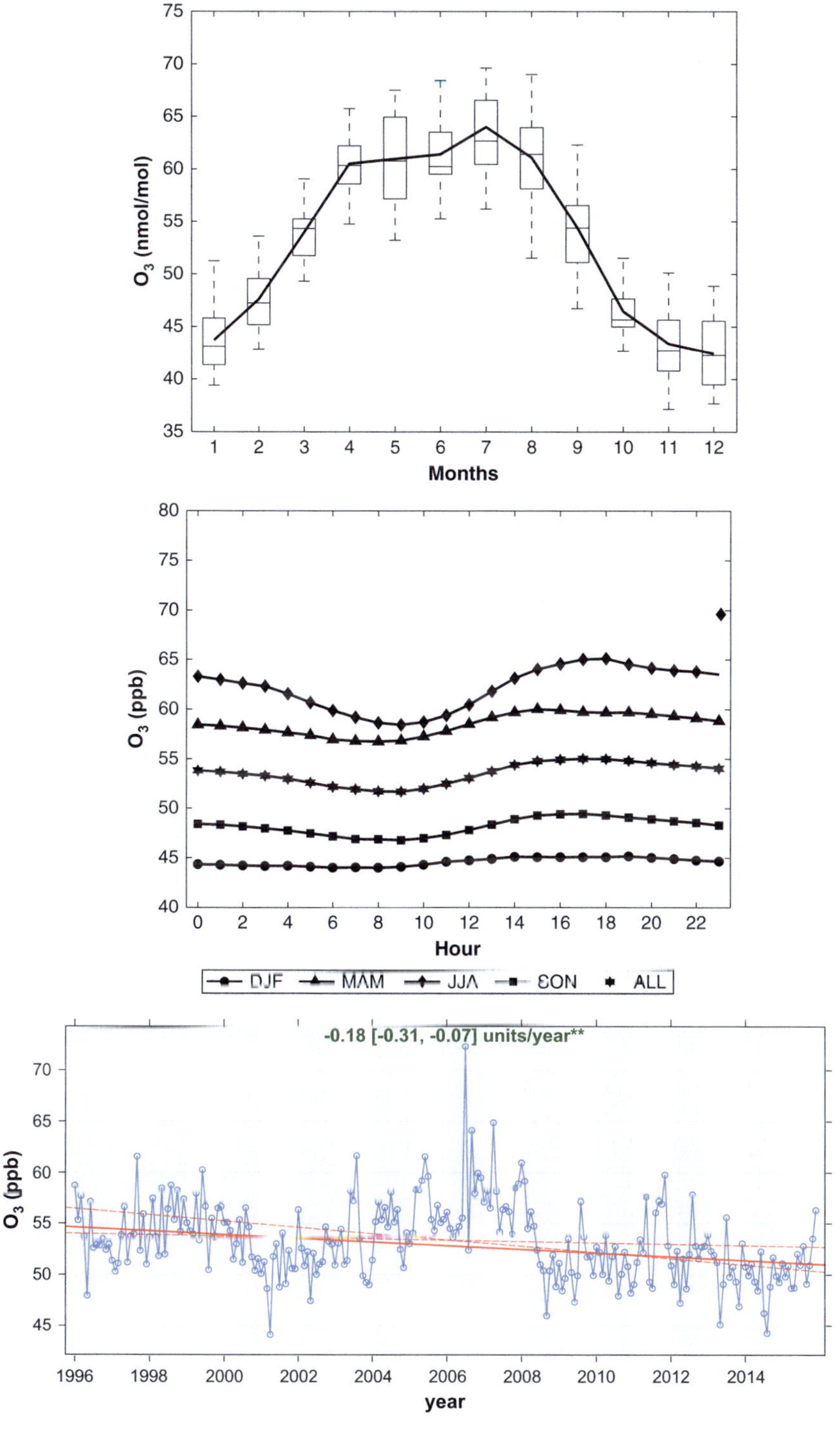

Fig. 5 Same as Fig. 1, for ozone (O_3)

2.6 Equivalent Black Carbon (Fig. 6)

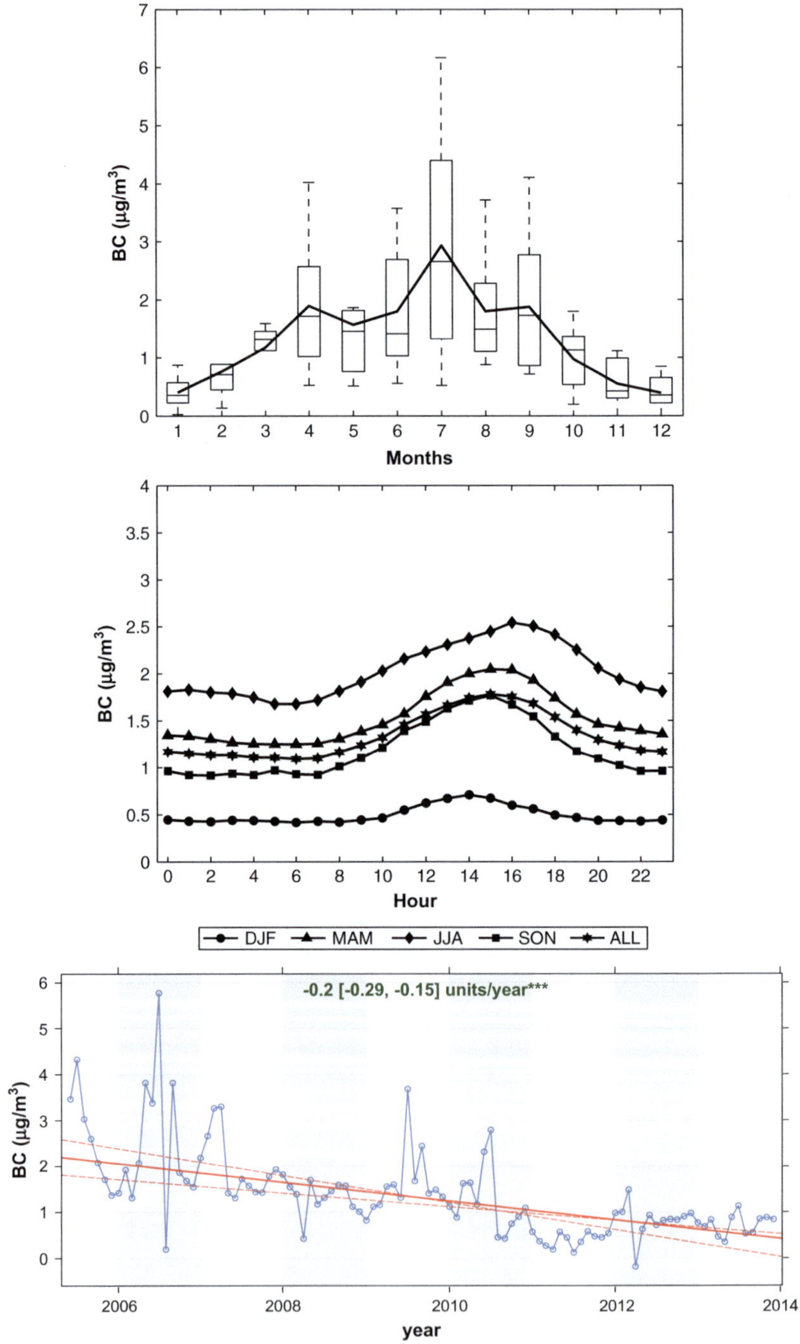

Fig. 6 Same as Fig. 1, for equivalent black carbon (BC)

2.7 *Accumulation (0.28 µm < Dp ≤1 µm) Aerosol Particles (Fig. 7)*

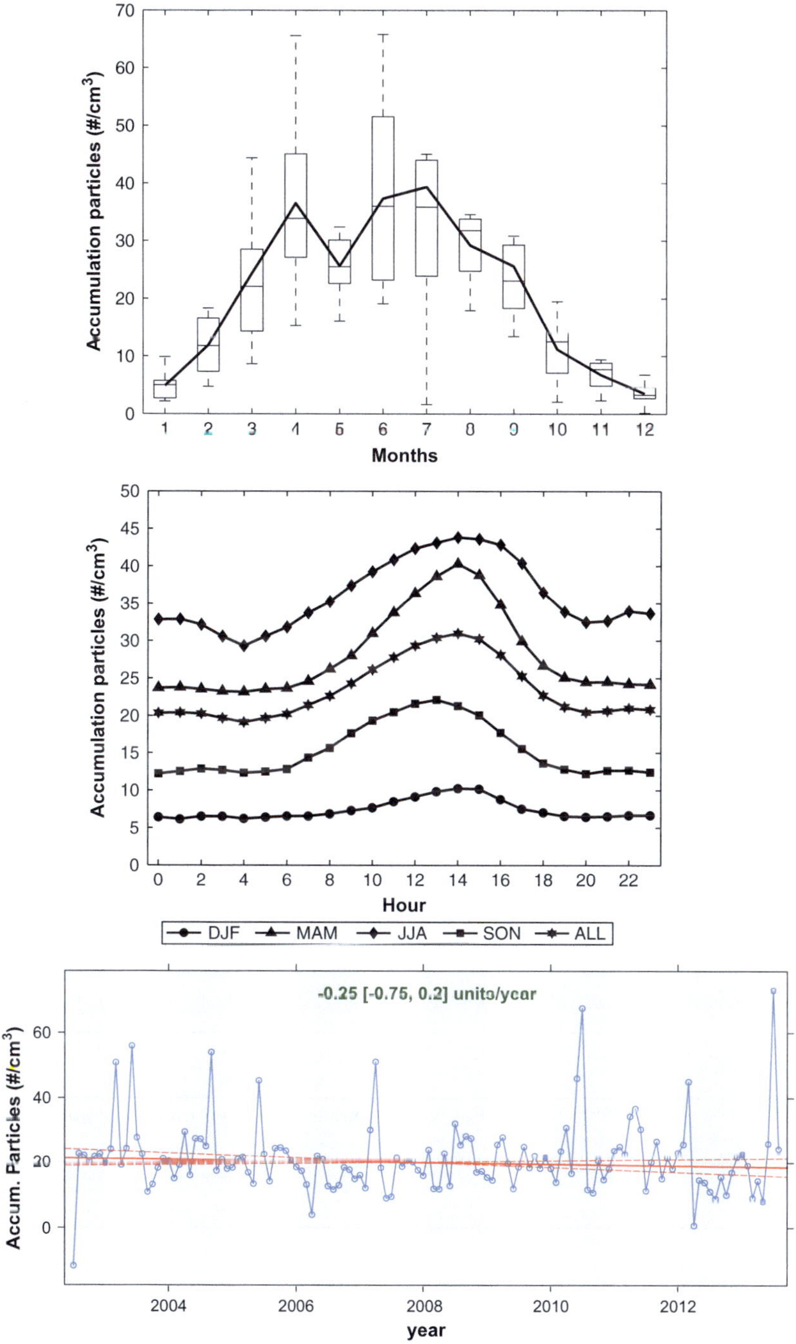

Fig. 7 Same as Fig. 1, for accumulation aerosol particles (i.e., 0.28 µm ≤ Dp < 1 µm, Dp being the geometric diameter of particles)

2.8 *Coarse (1 μm < Dp ≤10 μm) Aerosol Particles (Fig. 8)*

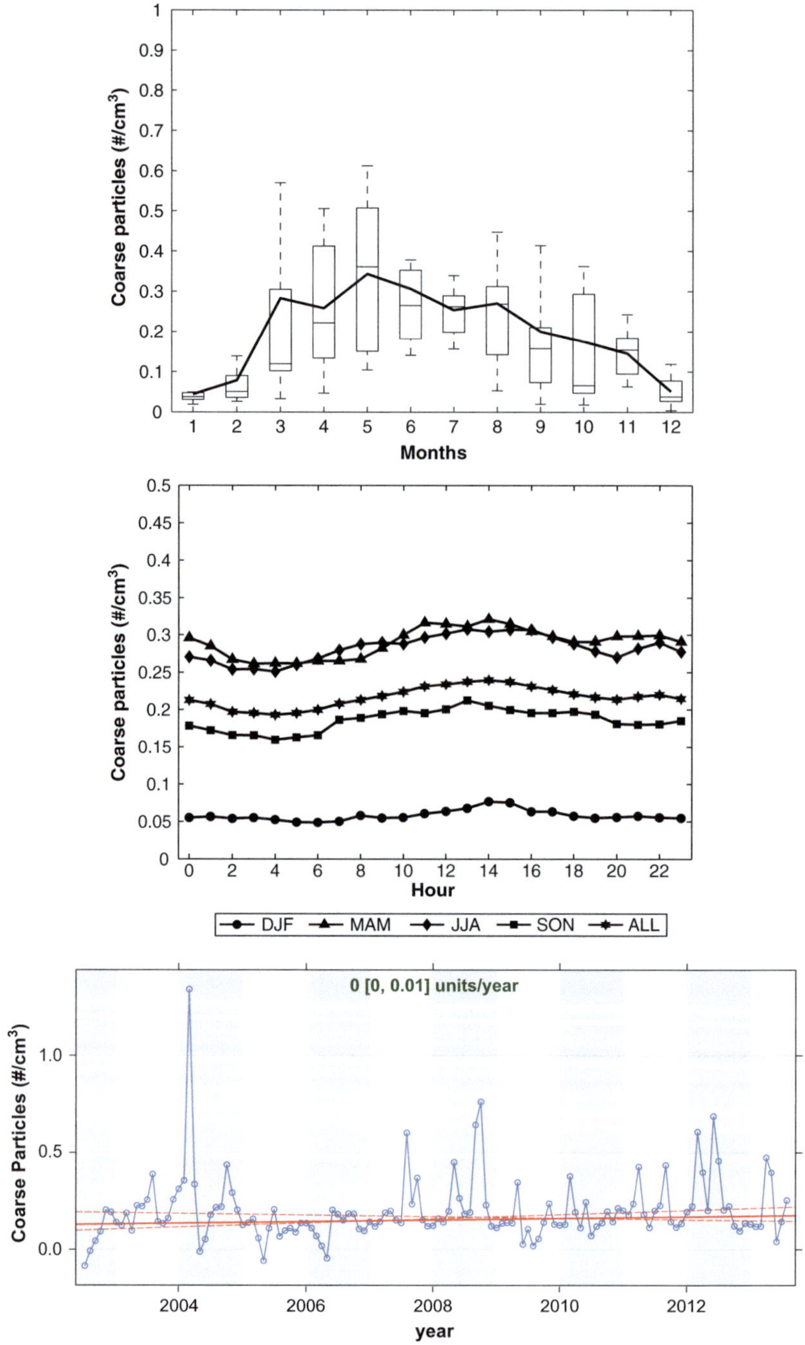

Fig. 8 Same as Fig. 1, for coarse aerosol particles (i.e., 1 μm ≤ Dp < 10 μm, Dp being the geometric diameter of particles)

2.9 Methane (Fig. 9)

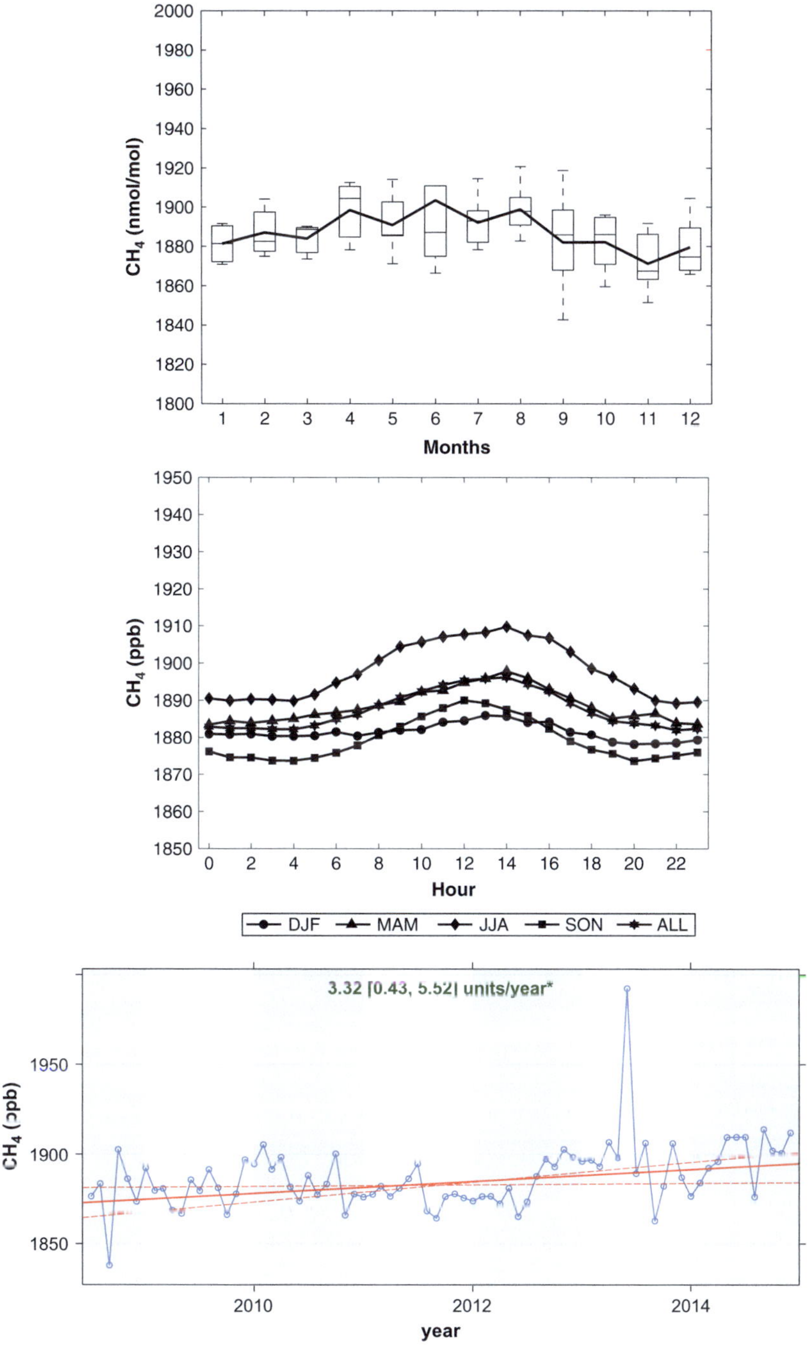

Fig. 9 Same as Fig. 1, for methane (CH_4)

2.10 Nitrous Oxide (Fig. 10)

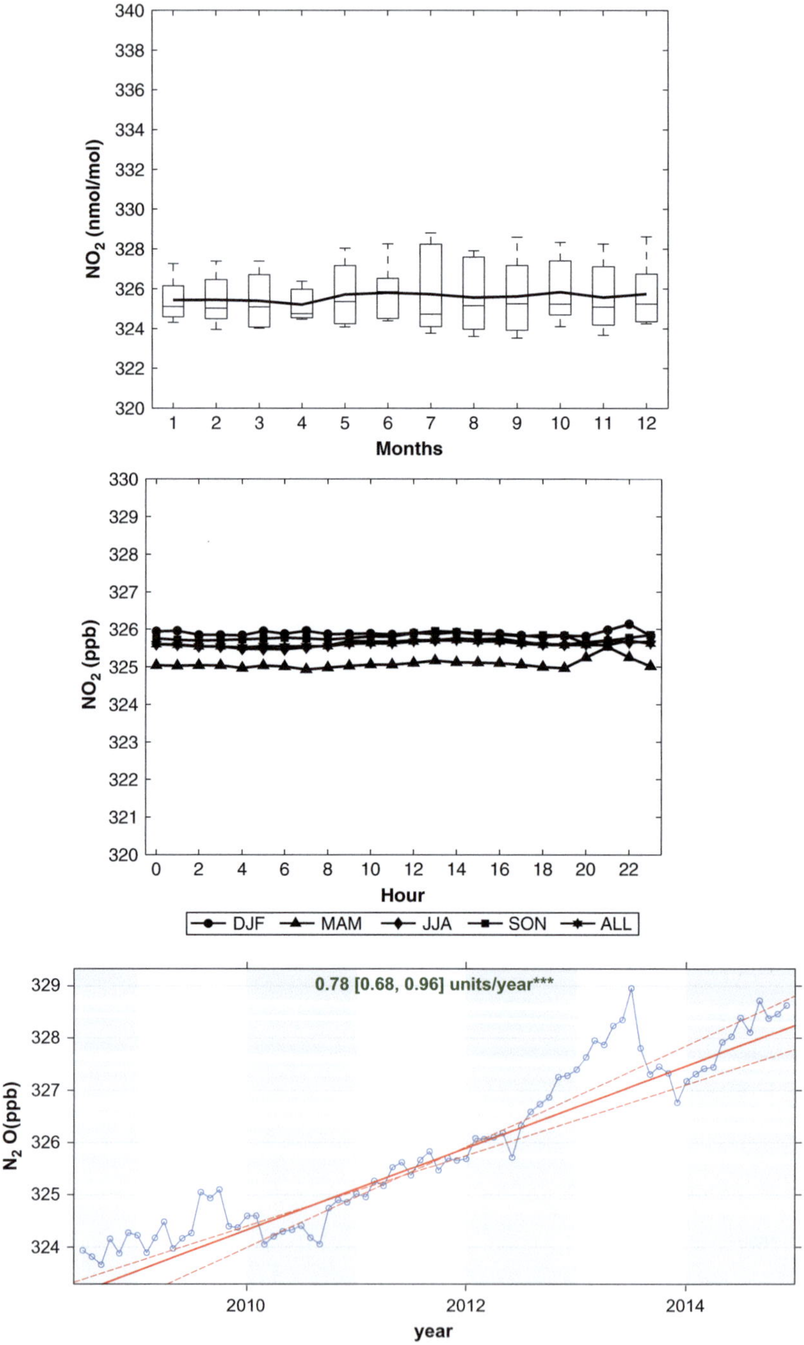

Fig. 10 Same as Fig. 1, for nitrous oxide (N_2O)

2.11 Sulfur Hexafluoride (Fig. 11)

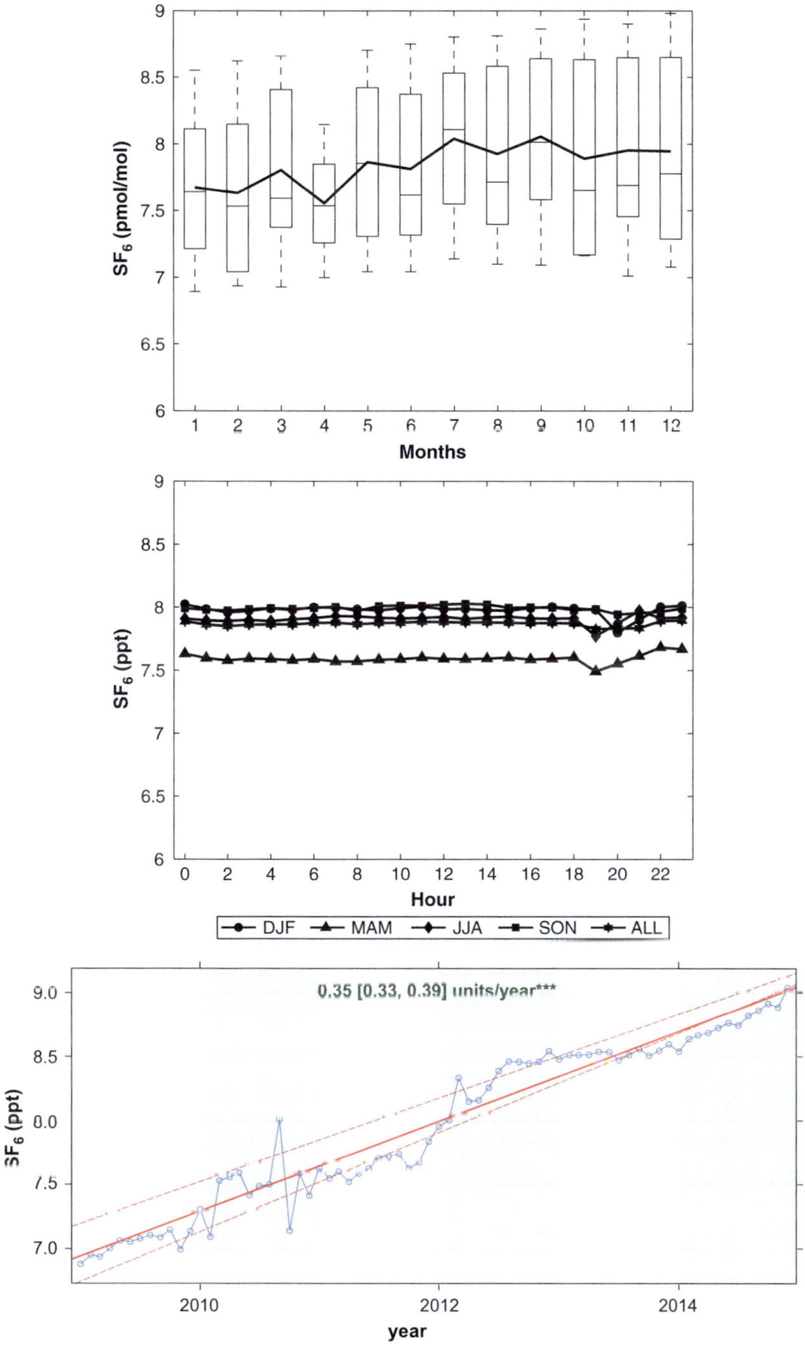

Fig. 11 Same as Fig. 1, for sulfur hexafluoride (SF_6)

2.12 *1,1,1,2-Tetrafluoroethane (HFC-134a) (Fig. 12)*

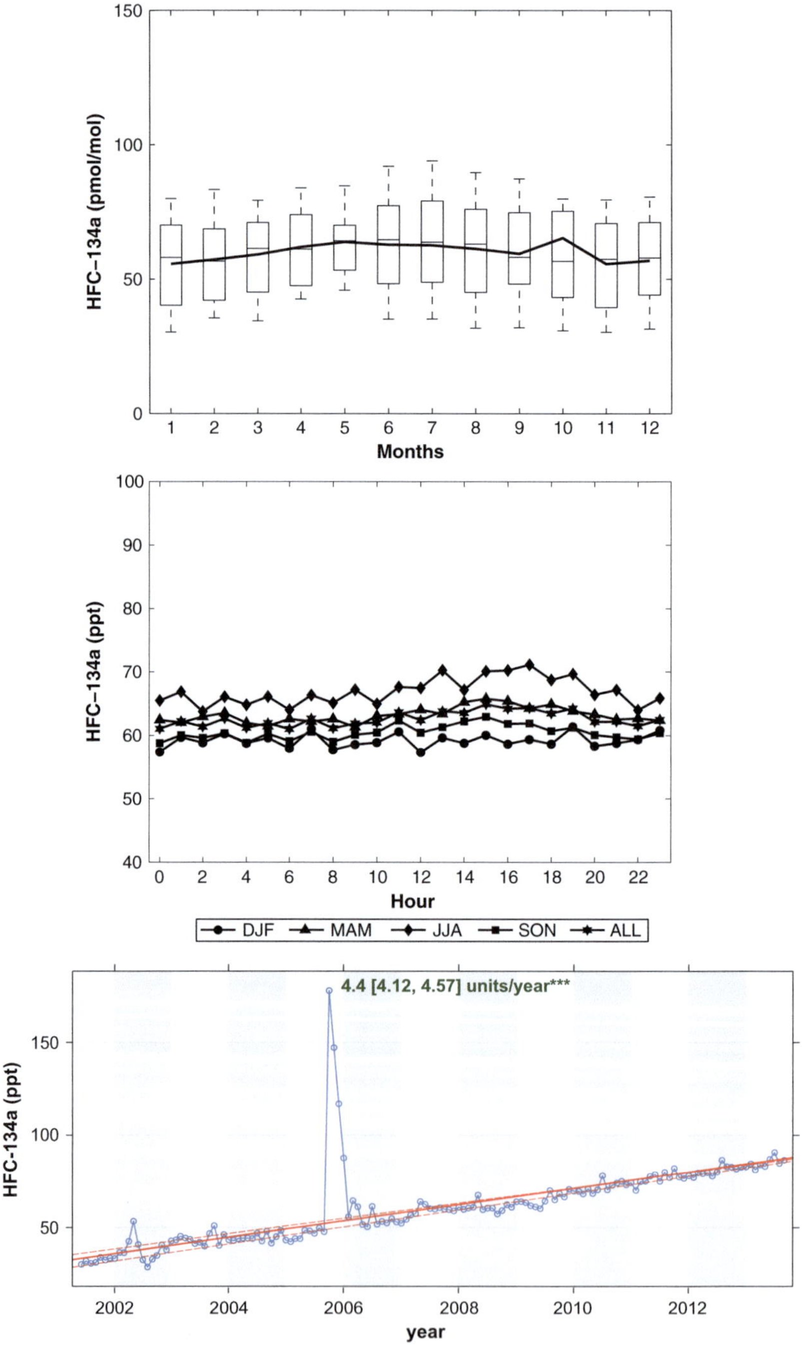

Fig. 12 Same as Fig. 1, for 1,1,1,2-Tetrafluoroethane (HFC-134a)

2.13 *Trichlorofluoromethane (CFC-11) (Fig. 13)*

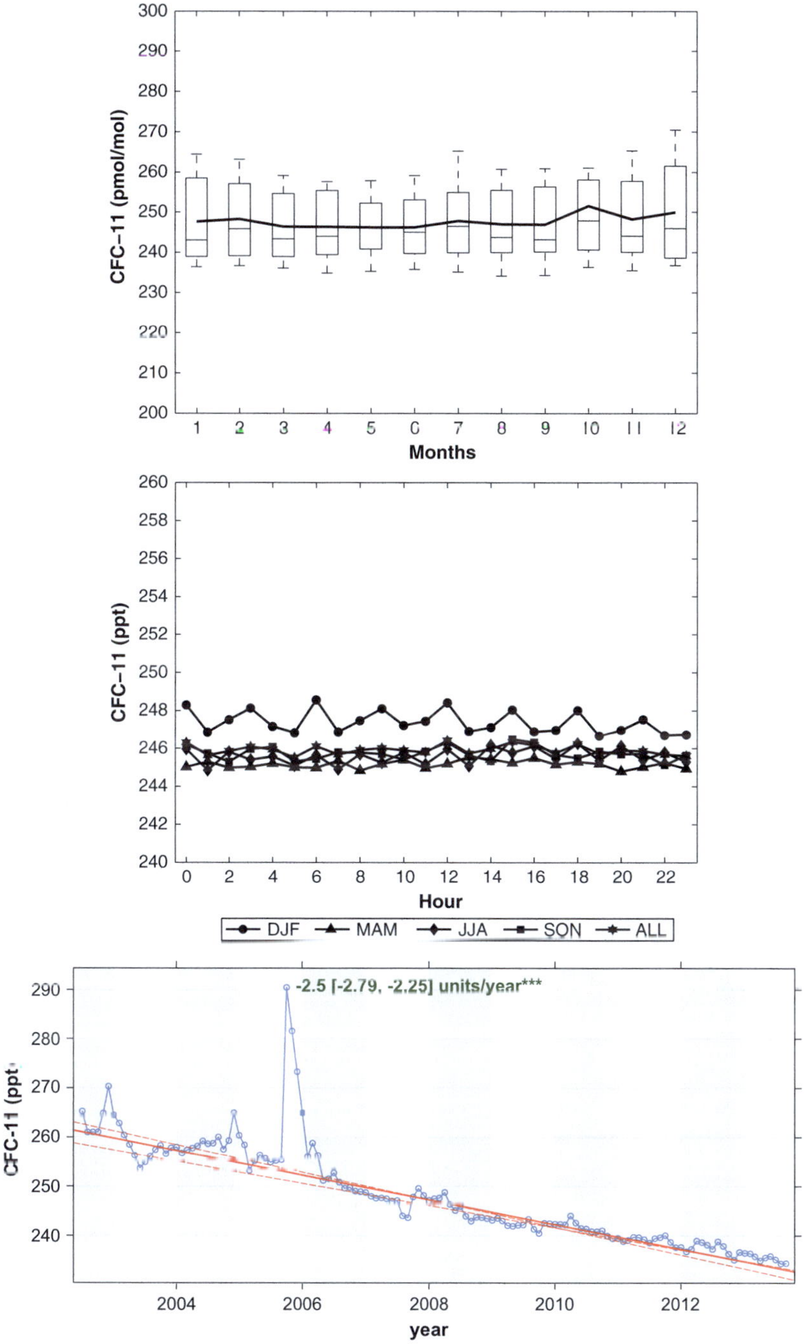

Fig. 13 Same as Fig. 1, for trichlorofluoromethane (CFC-11)

2.14 1,2-Dichlorotetrafluoroethane (CFC-114) (Fig. 14)

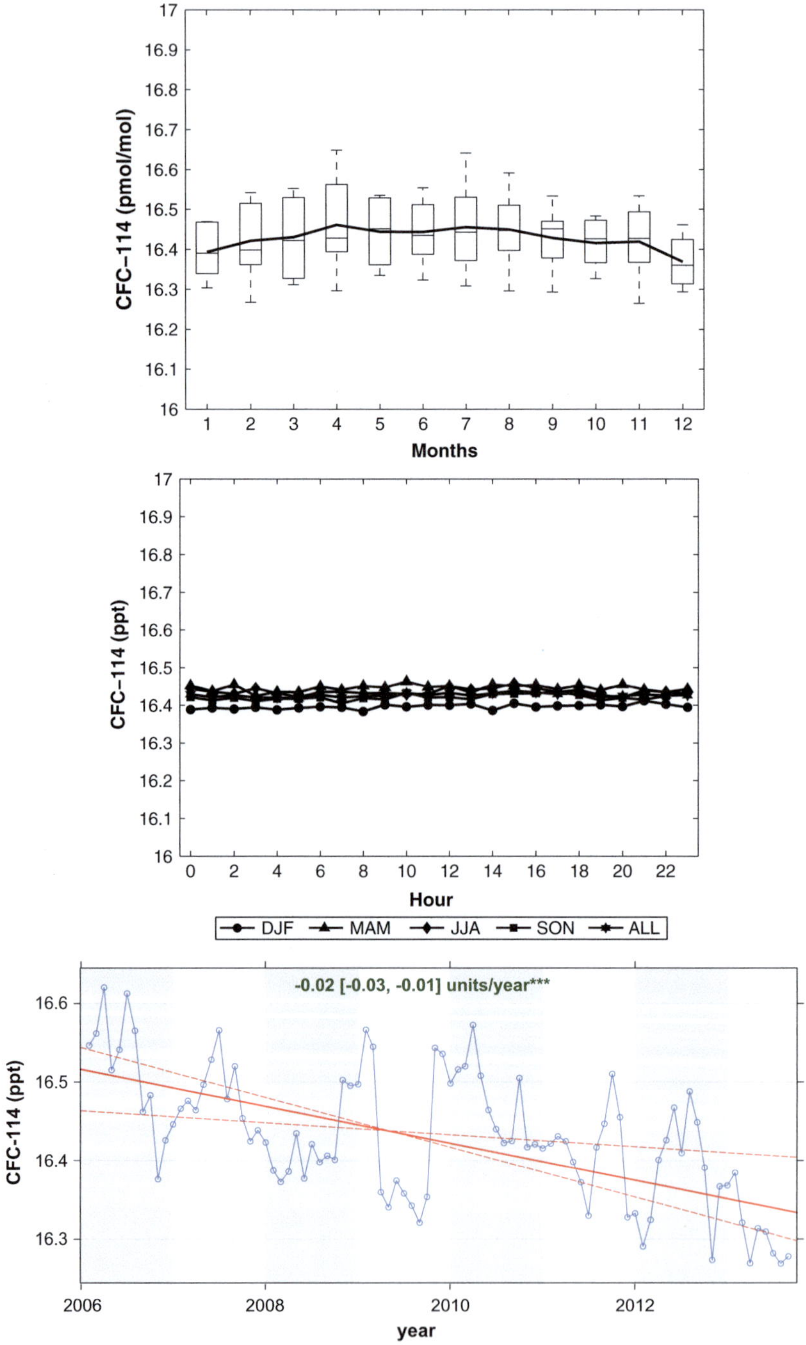

Fig. 14 Same as Fig. 1, for 1,2-Dichlorotetrafluoroethane (CFC-114)

Acknowledgments We would like to thank Dr. David Carslaw for providing to the scientific community the Openair R library (http://www.openair-project.org/), which has been used for the data analysis. Davide Putero grant was supported by the Project of National Interest NextData. Tony Christian Landi grant was funded by the scientific agreement between CNR-ISAC and ARPAE Emilia-Romagna.

References

Carslaw DC, Ropkins K (2012) Openair an R package for air quality data analysis. Environ Model Softw 27:52–61

Sen PK (1968) Estimates of the regression coefficient based on Kendall's tau. J Am Stat Assoc 63:1379–1389

Theil H (1950) A rank-invariant method of linear and polynomial regression analysis, Part 3. In: Proceedings of Koninklijke Nederlandse Akademie van Weinenschatpen A 53:1397–1412

MIX
Papier aus verantwortungsvollen Quellen
Paper from responsible sources
FSC® C105338
FSC
www.fsc.org

If you have any concerns about our products,
you can contact us on
ProductSafety@springernature.com

In case Publisher is established outside the EU,
the EU authorized representative is:
Springer Nature Customer Service Center GmbH
Europaplatz 3, 69115 Heidelberg, Germany

Printed by Libri Plureos GmbH
in Hamburg, Germany